Powder Puff Derby

Petticoat pilots and flying flappers

Mike Walker

WILEY

Published in the UK in 2003 by John Wiley & Sons Ltd, The Atrium, Southern Gate,
Chichester, West Sussex PO19 8SQ, England
Telephone (+44) 1243 779777

Email (for orders and customer service enquiries): cs-books@wiley.co.uk
Visit our Home Page on www.wileyeurope.com or www.wiley.com

Graphics by Dave Thompson, Thompson Graphics, Southsea, UK.

This publication is designed to provide accurate and authoritative information in regard
to the subject matter covered. It is sold on the understanding that the Publisher is not
engaged in rendering professional services. If professional advice or other expert
assistance is required, the services of a competent professional should be sought.

Other Wiley Editorial Offices

John Wiley & Sons Inc., 111 River Street, Hoboken, NJ 07030, USA

Jossey-Bass, 989 Market Street, San Francisco, CA 94103-1741, USA

Wiley-VCH Verlag GmbH, Boschstr. 12, D-69469 Weinheim, Germany

John Wiley & Sons Australia Ltd, 33 Park Road, Milton, Queensland 4064, Australia

John Wiley & Sons (Asia) Pte Ltd, 2 Clementi Loop #02-01, Jin Xing Distripark, Singapore
129809

John Wiley & Sons Canada Ltd, 22 Worcester Road, Etobicoke, Ontario, Canada M9W 1L1

Wiley also publishes its books in a variety of electronic formats. Some content that
appears in print may not be available in electronic books.

British Library Cataloguing in Publication Data

A catalogue record for this book is available from the British Library

ISBN 0-470-85140-6

Typeset by Dobbie Typesetting Ltd, Tavistock, Devon
Printed and bound in Great Britain by Antony Rowe Ltd, Guildford
This book is printed on acid-free paper responsibly manufactured from sustainable
forestry in which at least two trees are planted for each one used for paper production.

Bobbi Trout, pilot
1906–2003

The air up there in the clouds is very
pure and fine, bracing and delicious.
And why shouldn't it be – it's the
same the angels breathe.

Mark Twain, *Roughing It*

Contents

The Dames in Planes

Florence 'Pancho' Barnes, a cigar smoking ex-society girl, gained her nickname riding a mule through a bloody Mexican revolution disguised as a man. She slugged movie legend Erich von Stroheim hard enough to knock him down, fell in love with the Cisco Kid and flew across America to get him out of gaol. She was born on Pasadena's millionaire's row and would die outside a stone hut in the Mojave Desert. She was a woman who had more of 'the right stuff' than any dozen male pilots you'd care to mention.

Eugenie Shakhovskaya, the Russian princess who flew for the Tsar, sold secrets to the Germans and became a morphine addict and Bolshevik executioner during the revolution.

Dolly Shepherd, the London waitress who fell in with a sharp-shooting Buffalo Bill impersonator, ended up jumping out of balloons for a living.

Lady Mary Heath, a hard drinking, straight talking Anglo Irish aristocrat, put cowboy columnist Will Rogers off his lunch with her unguarded talk about women's sexuality. She was outrageous, fearless and a supporter of women's right to fly. Yet, despite her achievements, she died in poverty, alone, rejected by the society that had once idolised her.

Bobbi Trout earned the money for flying lessons by pumping gas but her confidence and flying skills knew no bounds – she took part in the world's first refuelling endurance flight made by women, staying aloft for 42 hours. She was to set numerous records and became wing-woman to her good buddy Pancho Barnes in the world's first women's flying corps.

Jessie 'Chubbie' Miller, the feisty Australian, persuaded a man she barely knew to take her on a world spanning flight as a mechanic, despite having no mechanical skills or ever having been in a plane! It was a flight that ended in a sensational love triangle and murder trial in Miami.

Hanna Reitsch wanted to be a flying missionary doctor but became a Nazi test pilot instead. She was the last person to fly into Berlin as Hitler crouched in his bunker and the Russians poured a million tonnes of high explosive onto the city.

Bessie Coleman, the first black woman to fly, gained her licence in the face of incredible odds.

Amelia Earhart, the golden girl of the air! Was she a great pilot or the puppet of her husband and publicist G. P. Putnam? He tried to control every aspect of her life, from the length of her hair to the length of her last, tragic, fight.

Amy Johnson took off from Croydon as an office girl who had quit her dull job to find adventure, and arrived in Australia as 'Amy, Wonderful Amy!' Yet, on her flight she carried a dark secret of suicide in the family and suffered from deep depressions. Her celebrity marriage to flying hero Jim Mollison swooped from triumph to tragedy and, during World War II, she was cruelly rejected by the establishment who did not want to use the skills of a 'stunt pilot.'

Introduction and Acknowledgements

There's a photograph of 19-year-old Katherine Stinson, the fourth woman to receive her pilot's licence in America, sitting in her first plane, a Wright B which cost $2000; no small sum in 1912. She is sitting beside her younger sister Marjorie, who was, at the age of 17, to become the youngest women to gain a pilot's licence. Both are smiling out from among the struts, wires and spars of an aeroplane only marginally more advanced than the Wright Flyer of 1906. Katherine, in open-necked shirt, trousers and buckled boots, holds a control lever in each hand; behind her is the petrol tank; above and below, the wings. The first thing you notice is how incredibly fragile the aeroplane looks – the idea of taking it up in any kind of wind seems suicidal to us now; it would be at the mercy of any sudden gust or thermal and in the face of lashing rain or an electric storm, surely the doped cloth, the wooden spars and the bracing wire would rip away in moments, leaving the pilot and passengers to fall to their deaths. And, often enough, that's what happened. Even the best, those who could 'sail the eighth ocean' as if born to it, knew that sooner or later there *would* come a situation beyond all their skill and experience and that then the fundamental magic of flight would have to be paid for . . . but no matter what the risks, and how many crashes they walked away from, there wasn't one of them who didn't think she had at least one more good flight in her.

This isn't a book about flying or a history of women in the air; there is history in it and there are women in the air, but mostly it's about pilots who flew when flying was still an adventure and most of the people who climbed into a cockpit wouldn't be allowed through passport control today. It's a story about characters who were driven by a desire to fly when a lot of people thought a woman on a bicycle was the fifth horsewoman of the apocalypse – about people who flew for the fun of it, because it was there, to prove a point, because it was a good way to meet girls, to earn a living, to get off the ground and simply because they wanted to, so why shouldn't they? Generally, I've stuck by the principle that a good story is always better than a good fact and stories probably tell more than facts about the way things were; as for how they were, the way they were told, I owe a lot of thanks to a lot of people. First of all, thank you to Bobbi Trout, who was there and flew the planes and made the records, who knew what it was like to fly in the dark above a dark country and who generously gave hours of her time to tell her own story and the stories of her friends, particularly Pancho Barnes, who flew alongside her in those days of wire and canvas. Thanks also to Ralph Barker, who not only allowed me unrestricted access to his extensive interviews with Chubbie Miller but shared his insights into the case and helped me to understand something of the character of a remarkable woman. Ann Tilbury-Harrington probably knows more about Lady Mary Heath than anyone else, and was extremely generous with time and material; she was also, as a long-time friend and correspondent of Elinor Smith and a historian of the period, able to provide many insights into people and their motives. Julian Temple, the curator of aviation at Brooklands Air Museum and Brooklands librarian Philip Clifford were unfailingly generous with their time and assistance, as was Paul Gladman, the archivist at Flight International. Balloonist Robin Batchelor pointed the way to information on early aeronauts; Anthony Doran and Isolde Walker provided invaluable help in tracing newspaper archives; Sylvia Lockett was a skilful and sensitive interviewer, and Sue Dean a mistress of logistics, so special thanks to Bobbi's babes and to Cheryl Baker for her knowledge and support. Thanks also go to Sarah Daniels and Annie Caulfield, just because.

Books are not just about life but also about other books, and any project of this nature is bound to rely upon the work of previous authors. I have given a list of sources at the end of this work but would like to acknowledge here the debt I owe to those writers who told some of the stories first.

The Powder Puff Derby

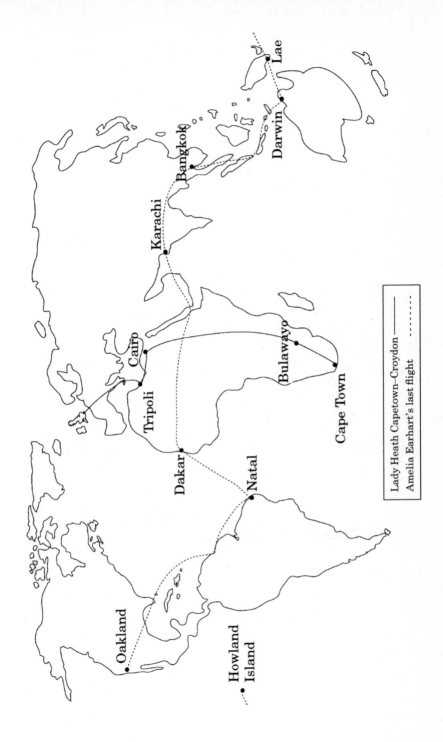

Lady Heath Capetown–Croydon ———
Amelia Earhart's last flight ·······

Oakland

Howland
Island

Natal

Dakar

Tripoli

Cairo

Bulawayo

Cape Town

Karachi

Bangkok

Darwin

Lae

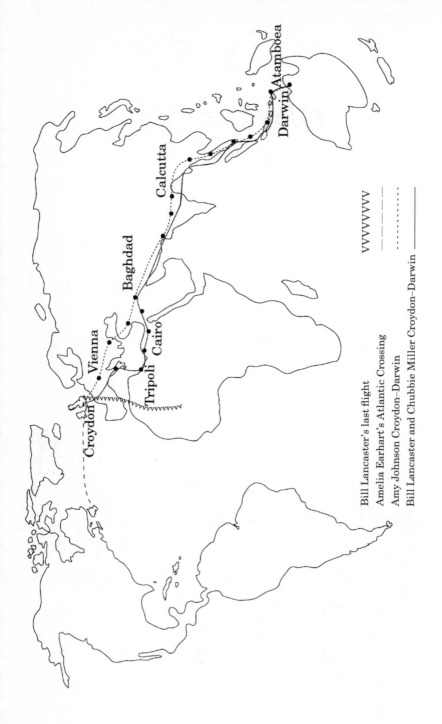

Darwin

Atambœa

Calcutta

Baghdad

Cairo

Vienna

Tripoli

Croydon

Bill Lancaster's last flight   VVVVVVV
Amelia Earhart's Atlantic Crossing   ————
Amy Johnson Croydon–Darwin   ·············
Bill Lancaster and Chubbie Miller Croydon–Darwin ————

1
Clover Field

For the thousands who had spent the night sleeping in their cars or under makeshift tents, the morning of 18 August 1929 promised a clear, warm day; a perfect day for flying. But then, as one harassed mechanic was heard to mutter to one of the horde of reporters wandering Clover Field and getting under everyone's feet, 'What other kind of day do we ever get in California?'

Those still on the Santa Monica highway had little trouble making their way out past the town to the airstrip; California's freeway system was still in its infancy, there were no traffic jams and the air had not yet begun to suffer from smog. In Ford model Ts and As, in trucks and sedans, the nation was on the move, as was its democratic right, looking for entertainment. The only clouds on the horizon were economic ones – the stock market had been falling drastically, thousands were losing their jobs as businesses crashed and President Herbert Hoover's government seemed incapable of halting the slide into economic depression. What's more, Americans were still being denied the right to walk into a bar and drown their sorrows: Prohibition, although thoroughly discredited and on its last legs, had not yet relinquished its improving grip on the nation.

What they did have was the spectacle of flight. It is almost impossible now to understand the power of that spectacle on the public imagination; the impact of the aviators who were, month by month, going further, higher and for longer than anyone else ever had. And each time they landed or crashed – though a Depression era joke suggested it was easier for a pilot to die of starvation than in an accident – the media was waiting to record the event. It might have been chance that the first great age of publicity tallied with the development of flying, but the coincidence did neither of them any harm at all.

Charles Lindbergh was one of the most recognised men alive. His achievement in crossing the Atlantic alone in his Ryan NYP *Spirit of St Louis* ensured him international fame – at least until his support for Nazi Germany became a little too common knowledge. England–Australia flyer Amy Johnson was the first great British popular heroine; she belonged not to the upper-middle classes but to the people, who called her Amy. Pilots like Wiley Post, who circled the world in eight and a half days in 1931, and Ross and Keith Smith, winners of the 1919 England–Australia race, were national heroes. The exploits of war-time flyers inspired novels, plays and films like *Dawn Patrol* and *Aces High*. In 1929 the Oscar for best production had gone to a flying movie: William Wellman's *Wings*, one of the last silent features to be released before the talkies flooded the nation's cinemas.

Drawing on these achievements, air racing had become one of the great spectator sports of the 1920s. This quest for speed was closely observed by other eyes too: In 1929 the US Department of the Navy had seen enough races to concur with the general opinion in the sport that engine cowlings significantly reduced drag and still allowed efficient air-cooling; in effect, they were getting their research and development costs met by the paying public. The Schneider sea-plane Trophy was first raced in 1914, then suspended during the war, only to be revived in the 1920s, when it inspired British designers to create and develop the Supermarine engine that would later power the Spitfire. The de Havilland Mosquito, one of the fastest fighter/bombers of the Second World War, was developed out of the Comet race plane flown by Amy Johnson and Jim Mollison.

The public didn't care about cowlings or the fighting potential of planes – they loved the romance and risk of long flights; the dawn take-off and the heart-stopping wait at the finish line: had the fuel lasted, was the engine holding out, what if the crew had been attacked on landing in some remote region or maybe the aviators had got lost? By no means an unlikely hazard in the 1920s, it was not uncommon for directions to be painted on roofs near airfields or over air routes to help those who had missed their way. Celestial navigation was not an accurate science, and there was always the weather: as Assen Jordanoff wrote in his pioneering manual, *Through The Overcast*, 'Learning to control an aircraft is a more or less simple matter, for plane and accessories are all man-made. But when you deal with the weather, which is neither man-made nor man-controlled, you are up against an entirely different problem.'

For those disinclined to wait the long days or weeks required by trans- or intercontinental fixtures, there were always the National Air Races, in which craft that were often no more than stubby wings and tails built around vast single engines with minimal cockpit space, competed over circular or figure-of-eight courses, with inevitable collisions and flaming crashes. The annual events were gladiatorial in their violence and attracted the flyers who had been unable to settle down after the Great War and had barnstormed their way across the country ever since. Pilots like Speed Holman with his black and gold Bullet Plane; the Air Hobo, a one-eyed buccaneer who designed his own aeroplane by drawing a chalk outline on a hangar floor and Granny Granville, who died when he deliberately crashed rather than hit a spectator who had wandered on to the landing strip. But pylon racing, for all its power and dash, did not have the glamour of the Air Derby, raced across oceans and continents by a wide variety of craft. In these contests, records were there to be broken, skill could still outwit engine size, and nature could upset both.

Among the spectators at Clover Field that August Sunday was Howard Hughes, engineer, aviator and movie producer, who had ideas of his own about record-breaking planes. By midday he was part of a crowd estimated to be in the region of 150 000, enjoying the side-shows and cotton candy, the ice-cream sodas and the funfair rides but all of them waiting for the main event – the start of the 1929 Air Derby. For this year it was a race with a difference; a difference not always appreciated by those who followed and wrote about the sport.

'I don't care what you guys write about their bravery, their skill, their sportsmanship or their adaptability to goddam aeroplanes. You can say what you like. But what I'm gonna say is, them women don't look good in pants.' With which, the unnamed journalist stalked off into the sun amid the sound of aeroplane engines being tested with a roar that was rising rapidly towards the unbearable.

'Dames in planes!' Worse: dames racing planes. Nineteen of them, setting out over a course of 2700 miles from Santa Monica, California to Cleveland, Ohio, where the eight-day National Air Races were running that year. The Derby winner was scheduled to arrive on day three, 26 August, after seven days of flying an average of 300 miles per leg, with stops for rest and refuelling.

The disgusted reporter was by no means alone in his opinion. There were many men, perhaps the majority, in the aviation world who were convinced that no one would arrive at all; that far from finding Cleveland, the girls would have trouble finding somewhere to powder their noses, and that the race would prove a shambles. Though far from equal in numbers – there were 34 women with flying certificates in America in 1929, as opposed to 4656 men – female pilots were determined that the race would not only be run but would prove a turning point as far as women pilots were concerned. It was an uphill struggle. In 1928, a pilot's licence cost $500. The average annual income for a woman was $800. It wasn't, then, so surprising if many women pilots were from wealthy, professional or showbusiness backgrounds and used to arguing their case and getting their own way. And one thing was clear – they were going to need every advantage they could get if the Derby was going to be flown the way they wanted.

Race favourite Pancho Barnes was inspecting her plane, smoking her usual stogie, spreading clouds of foul black smoke around, when one paterfamilias, passing with his brood, offered the opinion that if this was how lady pilots behaved, then none of his daughters would ever leave the ground. Pancho offered him one or two choice epithets and went back to her inspection. She, like most of her fellow women flyers, could see no reason why the sexes should not compete equally and this race, she reckoned, would be one good step on the way to equality in competition and prize money; and winning it, which she was convinced she could do, would enhance her own reputation as one of the fastest women in the world.

Pancho's buddy Bobbi Trout, who had earned the money to buy her aeroplane by working in a petrol station and was the holder of two height records already, was confident her flying ability would put her among the winners. She needed to place; when you didn't have your own money behind you like Pancho, publicity was everything. Fellow contestant Opal Kunze reckoned that power counted for something too: she had entered her Travel Air with a souped-up 300 hp Wright Whirlwind engine but had run foul of the race committee and been told it was 'too fast for a woman to fly'. They had their own very firm ideas about just how much engine power a woman could or should be allowed to have under her cowling. With a $25 000 prize at stake, Kunze hired and entered a less powerful craft. Pancho's Travel Air produced, as did the majority of the race planes, a more conventional 200 hp, though Edith Folz, the personal pilot of Don Alexander, chairman of the Alexander Aircraft Company, was flying the company's revolutionary Eaglerock Bullet, a low-winged monoplane with retractable undercarriage, enclosed cabin and a 250 hp punch. She too was considered a race favourite, though her careful style of flying would, it was reckoned, tell against her.

To provide real competition for those pilots with less power, the entrants had been separated into two classes – light sporting craft, like the Fleet Kinner flown by Australian Jessie 'Chubbie' Miller (yet to find herself on the front page of every newspaper in America as the scarlet woman at the centre of a love tragedy) or the De Havilland Gypsy Moth of German Thea Rasch, and heavier, commercial ships: the Travel Airs and the Lockheed Vega piloted by Amelia Earhart.

Before the race, Pancho gave an interview to the *Santa Monica Times*. The reporter asked, in the spirit of the age, how Mrs Barnes managed to balance her social duties with flying. Pancho told her it was no problem at all; flying was a real antidote to housework – in fact, to anything too conventional. Not that the aviatrix admitted to the reporter that being born was the first, and last, conventional thing she'd ever done in her whole life!

Two competitors who both felt they had a good chance of winning the trophy were also the youngest in the race. At 23, Louise Thaden, flying a Travel Air biplane, was already something of a celebrity. After receiving her pilot's licence, no. 74 signed by Orville Wright, at the age of 20, in June 1928 she set out to claim the women's altitude record, which she did, reaching 22 260 feet, with a makeshift oxygen tank. Two months later she lost it to her friend and rival, Marvel Crossen. At 25, Crossen had built up far more than the 100 hours flying time deemed a necessary qualification for entry to the Derby, and she had flown most of them in the inhospitable skies over Alaska, gaining invaluable experience of rough terrain and freak weather. After winning the race she intended to return and set up a commercial airline with her brother. Thaden, on the other hand, with a personal philosophy that espoused the principle, 'life stays a challenge as long as you look around and *find* the challenges', had started out as a saleswoman for Walter Beech of Travel Air and graduated to flying and testing company planes on her height and endurance record attempts. Beech had agreed that she should take part in the Derby, though, cannily he had also provided a specially-built craft for Marvel Crossen, reasoning that then he would have a better chance of one of his ships crossing the finish line first. As it turned out, the design of Thaden's open cockpit Travel Air was seriously compromised by exhaust problems with the Wright J5 engine and the young flyer was unknowingly facing a potentially lethal first stage.

Lady Mary Heath, who held the time record for the Cape Town–London flight, had entered her Avro Avian; the hard-drinking, straight-talking aristocrat who had put cowboy columnist Will Rogers off his lunch with her unguarded talk about women's sexuality was thought certain to place, then she had to withdraw for business reasons – but she promised she would be there at Cleveland, waiting to cheer the winner across the line.

By noon the temperature on Clover Field had climbed to the high 80s and the pilots were regretting the decision to hold back the start of the race until 2.00 p.m. The intention had been to allow the press and camera crews time to interview and photograph the pilots. This decision was in the hands of the race committee and although many of the women felt it should have been altered, the gentlemen had already eaten, as Pancho Barnes put it, 'so much crow', that it was agreed to let the late start stand.

The National Air Race Committee had dug in their collective heels and fought the women every step of the way. Badgered by the flyers to arrange a transcontinental event, they had suggested starting the race in Omaha rather than California, avoiding, by this route, the danger of crossing the Rocky Mountains. Amelia Earhart, who had achieved national fame as the first woman to cross the Atlantic as a passenger – a matter of considerable personal chagrin, since she did not consider riding alongside pilot Jack Stolz as worthy of anything at all – telegraphed the committee. It would be absurd to advertise the race as a serious event, she wrote, if the route avoided all danger. The committee considered the matter and agreed: the race would start from California but, just in case (of what, they didn't say), each plane would carry a male navigator/mechanic. For a while this decision stood, though no one really approved of the idea. Then one or two of the pilots got wind of a Hollywood plot. Certain producers were planning to enter glamorous starlets as 'aviatrixes' and have their planes piloted by the navigators and met by film crews at every stop. Once again, Amelia Earhart took it upon herself to telegraph the committee and point out how absurd this scenario would be in the eyes of the aviation world. She also stated that if she and her fellow pilots were not allowed to fly solo, they would not fly at all. Perhaps because of Lady Lindy's growing reputation – to her annoyance, a supposed resemblance to Lindbergh had prompted the nickname; she preferred to be known as AE – and certainly due to the fear of looking ridiculous to their fellow flyers, the committee once again backed down. The women would fly alone; there would be no film stars.

Except Ruth Elder, who didn't really count, since she had been a pilot before appearing in aeroplane films with silent stars like Hoot Gibson and Rex Dix. Flying a Swallow B5, Elder was determined not to be taken for a starlet. Before the Derby, her talents had mostly been seen on the silver screen and in advertising – she and Bobbi Trout had flown the inaugural flight for Pickwick Airlines, but only for the novelty effect. There was no chance that either of them – or any woman – would be taken on as a commercial pilot. Elder was determined to use her skills and her considerable charm to change this situation.

The planes were arranged in two lines across the field and, at 1.30 p.m., the celebrity starter, Will Rogers, known as America's favourite humorist for his laconic style, called for spectators and ground crew to clear the runway area as the race was about to begin. Being Will Rogers, he couldn't resist a speech. He was a long-time supporter of flying and had been promoting the upcoming event in his syndicated column for months, although his feelings about the proper place for a woman (he was a behind-the-kitchen-stove man on the whole) led him to categorise the flyers as 'petticoat pilots and flying flappers'. It was Rogers who had come up with a name for the race: the Powder Puff Derby, which had caught the public imagination and was to stick not only to the 1929 event but forever afterwards.

As the pilots strapped themselves in and checked their Rand McNally roadmaps and their emergency provisions in case of a crash landing – a gallon of drinking water, three days' supply of milk tablets, and a pack of beef jerky – Rogers eulogised the spirit of America's womenfolk. He described the zigzag route and quipped that if only Mexico City had been able to raise $50, they too could have seen the 'ladybirds'. The starting pistol was to be broadcast by radio from the National Air Races in Cleveland, though Rogers had a starter's flag. As the shot sounded almost 3000 miles away, he brought it down. The race was on!

2
Print the legend

No one ever would have said she was a beautiful woman – she wouldn't have said it of herself – but when she stood in front of you, the smoke from the black cigar jammed into the corner of her mouth stinging your eyes, the smell of gasoline and engine grease clinging to her as tightly as the oilslicked jeans she wore tucked into scuffed boots, then, somehow, as she lifted her aviator's glasses and measured your worth with a long, cool look, you knew she was going to get her own way. And you knew that her way was going to be outrageous, eccentric and almost certainly dangerous, if not damn near fatal; and there wasn't much you could do about it because she'd stood up to the toughest, this thickset ex-society girl who'd made herself into a flying legend.

She'd traded insults with Howard Hughes and drinks with Chuck Yeager; she'd pursued a simmering feud with Jacqueline Cochran, the good girl of women's aviation and fought the US Air Force to a stalemate; she'd spent her childhood in a vast mansion on Pasadena's Millionaires Row and would share her last days with coyotes and lizards and a few dozen half-wild dogs in a stone shack in the Mojave desert. She was born Florence Lowe but would live most of her life, after discarding an inconvenient first husband, as Pancho Barnes.

Her countryman Mark Twain might well have written that her life was 'full of surprises and adventures, and incongruities, and contradictions, and incredibilities, but they are all true, they all happened.' Twain had actually penned the words after a visit to Australia, but Pancho would never have let a mere island continent stand in the way of her life or legend; she would have hitched up her pants, wiped her hands more or less clean and growled: 'Hell, we had more fun in a week than those weenies had in a lifetime!'

And fun she had, like the day 'General' Pancho Barnes led her own air-force across the country on a mercy flight to rescue the Cisco Kid from jail – a mission that didn't quite pan out the way she had hoped, despite the personal intervention of President Roosevelt; or the night she told General James Doolittle of the Air Force, and authentic American hero, 'Damn it all, Jimmy, I can out-fly you and I can out-fuck you any time,' and went ahead, so she always said, and proved it.

Conventions existed to be undermined, overturned and thrown out, not that Pancho's sense of fun was invariably shared by her victims. About three months after gaining her pilot's licence in 1928 she formed The Pancho Barnes Flying Mystery Circus of the Air (there never was anything modest about her ambitions), reasoning that the best way to get experience and have some fun along the way was to fly like a madwoman and get paid for it. She joined up with a young parachutist called Slim Zaunmiller and they put together a series of stunts for the popular air shows that were becoming more common.

One of Pancho's favourite routines involved dropping a roll of toilet paper from the cockpit of the plane and following it down, flying through it, cutting it with her wings in a spiral so tight that every strut and wire was strained to the limit. After this, Slim would perform a series of novelty parachute jumps, finally swooping in over the gawping crowd and landing within feet of some bemused country girl who he would then proceed to charm into taking what was almost certainly her first flight.

She could see what a fine pilot Pancho was and Slim's easy manner soon allayed any fears. He even insisted that she wear a parachute so that nothing could possibly go wrong.

Installed in the open cockpit with Slim beside her, the passenger soon got over her anxieties and began to enjoy the unique experience of seeing the town she had lived in all her life from an entirely new point of view. Maybe she'd point out her house or the school she'd attended, or the shop or office where she worked, as Pancho brought the plane in low over the dusty streets or flew along the railway track while Slim explained how pilots used it as a guide across the vastness of the unmapped prairies.

Finally they returned to the field outside town where the air show was being held and took the plane up, higher and higher, until the spectators shrank to pinheads and the entranced young passenger could almost see the curve of the earth. And then Pancho looked over her shoulder with a grin and flipped the stick; the plane sideslipped dramatically, the cockpit spun towards the ground far below and in one heart-stopping moment Slim tipped the girl out, pulling her ripcord as she went. Pancho slammed the stick back, jerking the tail out of the parachute's path, leaving the shrieking passenger to fall through the emptiest space she was ever likely to experience – howling, flailing wildly, giving the crowd a gladiatorial, not to say burlesque, thrill as the parachute opened and, with its burden – limp now with shock and horror – drifted slowly down to earth where, after a lunatic descent, the plane had already landed and Slim was waiting to catch her before she crumpled to the grass.

The state of mind experienced by the novice parachutists can readily be imagined but, as far as we know, none of them were physically harmed, though whether their dignity recovered or they ever ventured above the first floor again in their lives is another matter. Certainly, nobody sued and no irate boyfriends, brothers or parents took a swing at the aviators. In the six months the Mystery Circus blazed its way across the skies of the American mid-west, Pancho was able to add 60 hours flying time to her logbook and Slim's name to her long list of lovers.

Sex, Pancho always said, was *almost* as much fun as piloting her own plane and she had no qualms about enjoying as much of both as she could pack into her busy schedule; however, while the men came and went, it was a love of flying that was written on her heart – an emotion inspired by the most important man in her life, her grandfather Thaddeus Lowe.

At her first air show, when she was eight and watching, entranced, as pilots like Glenn Curtiss and Lincoln Beechey flew low over the crowds in their experimental biplanes (and in 1910, just six years after the Wright brothers' first flight, *anything* that got off the ground and stayed up was experimental) her grandfather told her, 'When you grow up, everyone will be flying aeroplanes. You'll be a pilot too.'

The child had no reason to disbelieve him. He knew what he was talking about as far as the wide blue yonder went; Thaddeus Lowe had been one of the hot-air balloon pioneers of America, making his first ascent in 1857 at the age of 25.

Born in 1832, the child of farmers, Lowe had been apprenticed to a cobbler but his natural bent was towards scientific discovery and entrepreneurship, a fortunate combination, since the one would allow him to indulge the other. One of his earliest experiments involved flying the family cat by kite and, though it's hard to imagine what scientific principles he intended this to establish, he did discover that it was a popular spectacle among the kids of the town. It was a lesson the young scientist never forgot – no matter how fake or tawdry the show, grab the crowd's attention and they'll come back for more and pay for the privilege.

After finishing his apprenticeship, rather than settle down to mending other men's boots, he joined a carnival show and set up as The Professor of Chemistry, amazing and delighting the rubes with violent explosions and noxious smells conjured from mysterious coloured liquids and powders. Using the profits from his act he started to experiment with balloon flight, setting himself to learn everything he could about air currents and meteorological phenomena before delving into the chemistry of gases and the engineering of an airship large enough to carry passengers on continental journeys. It was a grandiose project and Thaddeus had neither the experience nor the finance to support the scheme. The airship refused to take off, his money began to run out, so he turned from the commonsense world of the carnival to the madness and mayhem of business, starting a company and soliciting funds to build a smaller craft which would test 'the mercantile and pecuniary benefits' of flight. Balloons were in the speculative air in those pre-war years and enough investors were persuaded to part with their money for Lowe to set up his company and begin his experiments.

His first series of tests was aimed at the upper atmosphere, where he concluded, prophetically, that high-speed winds would propel his craft with a swiftness that would make even transatlantic crossings viable. After one flight, when the winds were speedier than usual, he found himself seriously off-course and landed in South Carolina just days after the Confederacy had announced itself by firing on Fort Sumter. Descending as he appeared, from some 'etherial or infernal region', and quite obviously a Yankee spy, he was lucky to escape with his skin and balloon intact.

He always said it was the old carnival skills of persuasion that got him through, but when he arrived back in Washington, realising, as has many another entrepreneur, that war means opportunity, those same powers failed. He tried to convince the high command of the Union Army that this new war could be fought in the air as well as on the ground. The General Staff rejected the 'patently absurd' suggestion that balloons might have a military application; carrier pigeons were about as far as they could see – and they weren't too sure about them, figuring that hungry soldiers would shoot them down for lunch.

Not a man to be easily set back, Thaddeus arranged a demonstration of his ballooning skills from the lawn in front of the Smithsonian Institute, which had (so he maintained) recently awarded him the title of Professor for his lighter-than-air studies. Once aloft, the aviator sent a Morse code message to President Lincoln, informing him that the city, laid out below 'with its perimeter of fortified encampments', made a fine and bracing sight in the morning sun. Lincoln was less hidebound or more prescient than his commanders and it wasn't long before Lowe was given command of the newly created Union Balloon Corps, charged with reconnaissance and battlefield intelligence. Many years later, his granddaughter liked to say that Thaddeus Lowe was the real founder of the US Air Force, though he was never actually a member of the military. He operated as an

Irregular, dressed in frock coat and top hat, and would ascend in his metal-lined basket and telegraph information about the enemy's dispositions, take photographs and amend existing maps. He was under constant fire from both sides, since the average soldier wasn't quite sure what it was up there but he knew damn well he didn't like it. Lowe was once called 'the most shot-at man in the war' but he came through unscathed – though not unchanged.

After hostilities ceased, he decided to spend more time with his growing family, and settled in Pasadena where he concentrated on business and started up successful gas, electric and ice companies and founded the Citizens' Bank of Los Angeles. He built a spectacular narrow gauge railway along the spine of the San Gabriel Mountains, an impossible project that he brought to completion but that was to bring him, in 1899, to bankruptcy. However, if he had lost his money, he still had his dreams of a planet-spanning airship and a cable car that would cross mountain ranges. These dreams he shared with his grandchildren, and principally with Florence, a girl bought up to be genteel and determined (encouraged, one suspects) by the example of her unconventional grandfather, to be anything but!

Florence's childhood would have been ideal to anyone but Florence. Though grandfather Lowe had lost the family firm, her father, Thaddeus Jnr, had married a local heiress, Flora Mae Dobbins, and the family were able to maintain their position amongst the *haut bourgeoisie* of Pasadena. Tad Jnr raised thoroughbreds, whilst Flora Mae was a pillar of the local Episcopalian church of St James, an association that would cause her wayward daughter not a little inconvenience in times to come.

Mark Twain had typified the last decades of the nineteenth century as the gilded age, but there was nothing gilded about the Lowes – their wealth and lifestyle was 24-carat gold all the way through. The family ran a butler and they had servants, a swimming pool and a 35-room mansion, a half-mile diameter exercise ring for Tad Jnr's horses and for little Florence as well: she got her first pony at the age of three and was competing in local shows, and winning prizes, by the time she was six. She had everything she wanted – except her mother's attention. Her elder brother William was a weak and sensitive boy who took up most of Flora Mae's time and emotion (and made little impression on his sister or father) until his death from leukaemia when Florence was 12.

The girl, by then, was well set on her path in life and had discovered she preferred, and looked better in, jeans and shirt than in the frou-frou and big hats of late Edwardian style. Mother and child tried to come together in the face of William's loss, but somehow they just couldn't fit: Florence contemptuously kicking a box of French lingerie, a gift from her mother, across the room probably didn't help, but it does seem to have epitomised the relationship between the two women. With luck or wisdom beyond her years, the daughter realised she would never be a conventional girl; pretty in frocks, supportive in conversation, modest in behaviour. She was not and would not hurt herself by trying to be so: she would not be anything other than what she was, a decision underpinned by the wealth of her family but principally inspired by a temperament that loved risk and adventure and would try almost anything once.

After attending junior school quite happily – she was the only girl among 23 boys – she embarked on a career of wrecking whatever hope her mother had of turning her into a Pasadena debutante. She worked through four schools in eight years and ended up lying in a pool of her own blood on the floor of her dormitory, a suicide note pinned to her chest. Amid the screams of her room-mates she winked, grinned, got up and started cleaning off the red ink; it had all been . . . fun.

She did, however, have an ambition: she wanted to be a vet and work with animals, principally the horses and dogs, she loved. As far as Mrs Lowe was concerned, however, no way was any daughter of the Lowes of South Orange Grove Avenue going to tramp around inseminating cattle and gelding . . . No, the very thought was too much for her! Once young Florence had graduated, her mother put her foot down and persuaded her daughter to study art at a local college while a marriage was arranged.

The Lowes decided that the groom was to be the Reverend Rankin Barnes, the son of the incumbent at the family's very own and highly respectable Episcopalian church. He was youngish – only ten years older than his intended bride – and that was considered no problem; he was also handsome, had an easy manner and was popular with his parishioners. His position (he was soon to succeed his father to the pulpit of St James) would, Flora Mae hoped, lend him the authority necessary to curb his young wife's wilder ways. All in all, the Reverend Barnes seems to have been a pleasant, hard-working fellow, respected by his superiors, with a good future before him in the church; which rather begs the question, why did he do it?

He knew Florence; he had spoken at her college graduation ceremony and must have heard the rumours of her unconventional behaviour. Mrs Lowe had arranged early-morning rides for the young folks and while it is unlikely that the bride-to-be spat out any of her more colourful language, she would have been hard put to hide her essential nature. Maybe it was the prospect of sharing, and administering, as any decent husband might, the various large sums of money that would come to Florence from a number of trust funds that encouraged Reverend Barnes. On the other hand, after the marriage, Barnes never appeared to show any interest at all in Pancho's wealth, priding himself on supporting his family from his own stipend. It is more than possible that he was fond of the lively young woman – the couple were to remain friends long after they stopped being husband and wife – or perhaps he set a high value on an association with the family influence, and the Lowes and the Dobbins had much of that in Pasadena. Barnes never talked about his marriage in later years and although Florence talked about everything under the sun, she seems to have respected her first husband's reticence. Whatever the reasons, the Rev. Rankin Barnes was ready and willing to let himself in for what he was surely going to receive.

The real mystery lies inside the 18-year-old Florence Lowe. What in hell's name, as she would have said, was she thinking about? This girl had quite calmly led her favourite horse up the oak-panelled stairs of her school and installed him in her room, explaining to an enraged headmistress that, 'Poor Dobbins was so lonesome he came to find me.' Something, one might speculate, that her busy mother, Flora Mae Dobbins Lowe, would never have done.

Lauren Kessler, in her Barnes biography *The Happy Bottom Riding Club*, suggests that Florence was still, despite all, seeking her mother's approval and this, combined with the attraction of getting out from under the parental roof at a time when it was difficult for a young woman to commit herself to a career, was enough to push her in the required direction. Besides, the Reverend Barnes was by no means an unappealing fellow. The prospect of being the centre of attention at a big society wedding probably played a part, as did a natural pride in getting one over her prettier school friends by marrying the most eligible bachelor in town. When Barnes had given the graduation speech at her last college, he had caused a number of hearts to flutter though, significantly, not Florence's. And, after all, if there was no real reason *to* marry, there was equally no reason *not to* and though it would have been fascinating to know what advice grandfather Thaddeus Lowe might have given, he was long gone by now and, unless she were to jump aboard one of his Planet Airships and fly out to the territories, it looked pretty much like Miss Lowe was heading for the altar and the marriage bed.

Out of the two, the altar was the better option. The Lowes were an important family, the wedding was all that could have been expected: the church decked out with greenery; carriages and gentlemen in morning dress; the bride in organza ruffles with a bouquet of roses and lilies; the groom's father officiating, as solid and trustworthy as the church which stood, four-square and Protestant, around them. Back at the Lowe mansion a wedding feast awaited the guests and while an orchestra played popular tunes, the bride changed out of her gown and into an expertly tailored suit that brought out her best features.

The young couple left by limousine for the first night of their honeymoon in San Bernadino; on arrival they walked in the limpid evening light, ate dinner and made their way, together, to a marriage bed that neither regarded with anything short of extreme unease. For all her wild ways, Florence Lowe's ideas of love, marriage and the duties thereof were learned from romantic fiction and films. Her new husband, older by a decade, was her equal in inexperience; his ideas stopped and started with the bits you could see – chatter across the breakfast table, the minister and his wife entertaining the sewing circle or opening civic functions, and, of course, church on Sunday – the less mentionable functions, apart from the necessity of conceiving children, had best remain exactly that: unmentionable.

The first night was a non-event. Florence lay on her side of the bed in her new, lacy nightgown; Rankin lay on his, pyjamas buttoned up to the neck: neither made a romantic move. The next morning they travelled on to the Grand Canyon where the grandeur of nature had little or no effect on the heights, or depths, of mutual passion and the second night followed the same pattern as the first. On the third night, Rankin Barnes pulled himself together, looked his conscience firmly in the eye and made it clear that, distasteful as they both found the subject, it was now his wife's duty to provide him with his conjugal rights. Given her later appetite for, and open delight in sex, Florence's first experience could not have been more unpromising. She reported later that neither of them enjoyed it; not that Rankin expected to, it was all down to duty as far as he was concerned. Florence, who had hoped for more, hated it from start to finish and decided there and then that there would be no repetitions; as far as we know, she and her husband only spent that one night together throughout their marriage and for a woman whose luck was to hold remarkably good far longer than she had any right to expect, it must have seemed like a bad joke when, within weeks, she discovered she was pregnant.

She didn't like that much either – though characteristically she boasted to her friends: 'Guess what, I gotta bun in my oven. That's more than you bastards can manage!' A few months after the couple had moved into their new home, she began refusing to go out, citing her awful, unnatural shape. She did keep riding her beloved horses as long as she was able to, but married life was about to provide her with another shock: she was expected to behave like the minister's lady.

Not unreasonably, the Rev. Barnes intended to support his wife on his income and, in return, receive her support for his ministry. He was a popular preacher, at least with the local ladies, and was often called out on late-night missions of spiritual mercy. A beacon of propriety, he always took Florence along too, though she had to sit in the car just in case Barnes had to use the excuse of her presence to make a quick getaway. She was also expected to teach Sunday school, a task she got round by bribing her pupils with jack-knives for learning their texts.

Being pregnant, stuck in a respectable household where homily was the order of the day and, for the first time in her life, without the resource of unlimited money, might well have made Florence think deeply about her life and position. This might well have happened with any other young woman but not the new Mrs Barnes. She was, and always had been, a woman of action. First, however, there was the matter of the baby.

Taken into hospital for the birth, she insisted that her favourite dog, Nicky, a ferocious German Shepherd, accompany her; the doctors, of course, refused outright. The dog moved in under her hospital bed and stayed there, growling at everyone who got closer than a couple of yards. After the birth (the baby was named William Emmet, after her dead brother), Florence brought him home, laid him on the floor so Nicky could have a good sniff, and more or less forgot about him for the next 14 years, consigning him to her husband and a series of nurses.

Within a couple of years of the birth of her son, Florence's mother, Flora Mae, died of a heart attack – high blood pressure ran in the family and was to cause serious problems for Florence later in her life. Once the funeral was over, it was clear that apart from the Rev. Rankin Barnes, there was no one left to please any more by trying to act like a conventional young wife.

Florence's first concern was to get away from the stifling atmosphere of life at the rectory, which led naturally to her second concern: the means of escape. She didn't want to divorce her husband: for all their incompatibility, the two weren't bad friends, just hopeless partners, and they reasoned, without exactly stating the facts, that separate lives under the same roof was the best answer for them both. It would give Florence the status and freedom of a married woman and allow Rankin to continue his ascent into the hierarchy of the Episcopalian church unchecked by scandal.

Various bequests beckoned, but they were all in the future and right now Florence needed money to pay for a full-time nurse for little Billy and cover the expenses that the housekeeping did not, which meant just about everything she really wanted to do. Thinking the matter over, she came to the inevitable conclusion that the one saleable talent she possessed was raising, training and working horses, and she began riding in public again, attending shows and exhibitions, picking up what jobs she could, including a stint as a double for the travelling evangelist Aimee Semple Macpherson in the preaching and riding act she took around local rodeos.

Better paid and more exciting work came her way in the film industry. Dozens of cowboy films and serials were being shot in the hills and scrubland around nearby Los Angeles, and Florence began to gain a reputation as a horse wrangler who could get her animals to do just about anything the director required of them, from running alongside a careering stagecoach to making death-defying leaps over yawning crevasses so the hero could escape the posse or the Indians or the men in the black hats. She could also, in a time when union influence was less powerful than it later became, heft the weight of a camera for mounted tracking shots or turn her talents to continuity, lighting a set or doubling for the star, and with her stocky build she could cover for hero and heroine alike, a considerable saving on the usual tight film budgets. She even tried her hand at script-writing, renewing a friendship with the Austrian director Erich von Stroheim, whom she had met years before when he was working as a stable boy while trying to get a foothold in the film industry. They started a number of screenplays and set up a company to novelise popular films, though the venture came to nothing since both soon moved on to other projects.

By the winter of 1924, Florence was able to use her film money to hire a nurse for Billy, a couple of housemaids for the rectory and have a good time for herself. It was the Jazz Age, and she began running with a young and dangerous set, half film people, half wild young society kids, all of them determined to prove that Prohibition was not going to stop them having a great time. Rankin Barnes began to find himself lumbered with the kind of house guests who usually lurked in his worst and wildest dreams; he also found himself with bathtubs full of brewing gin and midnight visits from the local bootlegger. This was not the way to impress the good people of Pasadena, and he asked Florence to shift her activities elsewhere. She was happy to do so, moving into a cliffside mansion she had recently inherited.

The couple still got together at weekends, after church, which Florence did not attend (there was no young wife looking up adoringly at Rankin as he preached), when they would take baby Billy over for lunch with Florence's father and his new wife – until the baby's vile and utterly uncontrolled behaviour drove Tad Jnr to request an end to familial visiting.

According to some accounts, at this time Florence began an affair with a student at Pomona College, a young man as eager as she was to 'discover the mysteries', as she put it, and having discovered them once, the young couple proceeded to spend the next few months discovering them all over again and again, until Florence, feeling like she'd 'won a battle and whipped the world', was unable to resist sharing her happiness with her husband. And one has to say it, there was definitely something of the saint about the Rev. Barnes, because he forgave her and merely requested, though in his firmest voice, that the affair cease. Florence, having just come into an inheritance through her late mother, agreed to leave her boyfriend and the country on an extended cruise down to South America.

On the cruise she was to meet the first of a series of lovers who conformed to her ideal of a real man. Don Rockwell was a tall, tanned, devil-may-care adventurer with dark eyes and a dazzling smile. He was sophisticated, a man of many worlds, not all of them strictly legal, he was a good and thoughtful lover and made it quite clear from the beginning that he had no intention of ever being caught in the marriage trap. He was the hero of a thousand romantic novels and films, the man Florence had married in her dreams and, by rights, he ought to have been a lothario who would use and abuse the young woman who'd fallen for him and leave her lovelorn and alone. Not that Florence was exactly a wilting flower – she had no illusions about Don, he had none about her, they enjoyed each other with gusto as the liner cruised down to Rio. Don even went so far as to dedicate a poem, *The Jungle Kitten*, to her 'tawny, satin hide' and penchant for being 'fondled all the time'. The romance lasted the length of the voyage and ended as the voyage ended – with a brief final act in Greenwich village. Florence had come a long way in a few years, though once she was back in Pasadena it didn't seem all that far at all, and the Jungle Kitten began looking for something else to test her mettle.

One drunken evening, sitting around with a group of actors, stuntmen and cameramen, sharing wild stories – and in America during Prohibition anyone who took a drink could find themselves on the wrong side of the law – one of the party suggested they get jobs as crew on a freighter he knew of that was heading down to Mexico. It sounded fun, maybe a bit risky, but he assured them the ship was seaworthy and the Captain at least as sober as they were. As it turned out, the Captain showed a good deal of common sense when presented with the idea and insisted on hiring a friend of his, Roger Chute, a graduate with wanderlust and sailing skills, as helmsman for the trip.

Florence figured that no one would let a girl aboard a tramp steamer so she cut her hair, slicked what was left back with oil, slipped into jeans and a shirt and, with a cigar clenched between her teeth, was able – or allowed on account of her money – to pass as one of the boys. With Chute navigating, the voyage passed comfortably enough, the 'crew' slipping ashore every so often for a wild night, until they reached the Mexican port of San Blas, where they found themselves on the edge of a revolution.

The *Cristeros*, religious rebels, were fighting a war against what they saw as the anti-Catholic policies of the Mexican government; the army and the forces of law were in the centre of the country facing the fanatical insurgents, leaving the coast undefended against bandits, a small army of whom were besieging San Blas, shooting at random into the town from the surrounding hills, demanding a vast ransom before they would pack up and leave. The sight of an American ship in harbour offered a solution to the townspeople: they boarded the vessel, impounded it, then loaded all their valuables on to it. The bandits were determined that the vessel would not escape with the booty. In some stories, the ship itself was involved in the rebellion, carrying contraband arms for the *Cristeros*. True or not, the outcome was stalemate: The Captain wouldn't or couldn't put to sea, the townspeople wouldn't or couldn't get off, so the days passed as everyone sat it out under the bandits' bullets.

Until one dark night when Florence, or Jake Crow as she was known aboard, took a late walk on deck and saw Roger Chute, the helmsman, preparing to slip overboard. He'd had enough of sitting around and was going to make a break for it, heading inland before turning back towards the US border. 'Jake' insisted on going with him, but he wasn't fooled; he had no intention of being stuck with a girl on this dangerous trek. This was not the best thing to say to Florence at any time and after she'd proved herself as a deckhand for several weeks, working and drinking as hard as any of the guys, it was foolish. If he was going, she was going. Rather than give up the chance, Chute agreed, she could come along but if she fell behind, she'd be on her own.

They slipped ashore, bribed their way through the bandits' lines with whisky, hired a horse for Chute and a mule for Florence and set off across country. It was the hot season and the land was brutal, with mile after mile of desert, sage, cactus and blinding sun out of which bandit, soldier or revolutionary might appear at any moment; and Florence was having more fun than she'd ever had before in her life. Looking at Chute, sitting atop his skinny white horse, she told him (in one version of the story) he looked like Don Quixote, setting out on his adventures. Chute, who considered himself to be something of a philosopher, responded that if he was the Don, then Florence was surely his squire, Sancho Panza. Florence said she preferred Pancho to Sancho and Chute agreed: it was a good Mexican name and for the duration of the trip that's who she'd be. He didn't for a moment guess that, from that time on, for the rest of her life, that's who she'd be: Pancho Barnes.

Chute's misgivings about Pancho's stamina were to prove misplaced; as they trekked over a thousand miles to the eastern seaport of Vera Cruz, she not only kept up, she kept him alive when he contracted blood poisoning and she had to commandeer a passing American so the two of them could stretcher him miles across rough country to an oil refinery doctor. In return, Chute shared his philosophy of personal responsibility, toughness in the face of adversity and enjoyment of the moment. This sounded pretty good to Pancho, who'd just crossed Mexico, been fired on by rebels and government troops, begged, borrowed and, on occasion, stolen food and drink and all without spending a cent. She was beginning to see what she really wanted out of life: adventure, sex and good buddies who'd risk their lives for you, just as you would for them.

Shortly after arriving back in Pasadena – where she called on the Rev. Barnes, brown as a nut, poncho-clad with a cigarillo between her teeth – she was to find the activity that would give her the adventure she craved. She and Rankin Barnes also finally separated, amicably and still without thought of divorce. The final straw for Rankin had come shortly after Pancho's return home, when she mentioned casually (or not) at a dinner party one night, that she was feeling antsy and really needed a good fuck. He hastily applied for an executive post at the New York headquarters of the church; she went out to a nearby airstrip with her cousin Dean, who was going to take flying lessons.

Pancho, being Pancho, decided that flying looked like fun and asked Dean's teacher, Ben Caitlin, if she could sign on as a pupil. Like a lot of pilots, Caitlin had little time for women out of the home and none at all for them in the cockpit. He explained that he had reluctantly taken on three or four girls as pupils in the past and none of them had soloed or even finished the course; they just didn't have the grit to do the hard work required. Looking at the ramshackle operation Caitlin was running – paint peeling off his hangar, office and run-down Travel Air biplane – Pancho reckoned ready money might be an inducement and offered to pay for a trial flight there and then. Caitlin reckoned that five dollars for fifteen minutes was fair – and would at least buy a drink or two for the boys that evening – and agreed to take her up. He thought that if he threw the plane around the sky violently enough, she'd soon have all and more than she wanted of flying and he'd be rid of her. He was wrong, of course.

After loops, turns, spirals, stalls, dives, spins and rolls, after 'wringing her out' as pilots put it, he landed and asked if she still wanted to learn to fly. 'Hell, yes,' she said, 'I want to learn to fly.'

Over the next three months Pancho dedicated every free moment to Ben Caitlin and his Travel Air, learning the hand signals by which pilot and pupil had to communicate, becoming familiar with simple but basic manoeuvres like banking and figure-of-eight turns, getting used to landing and taking off and 'touch and goes', where the pilot just allowed the wheels to kiss the field; she also had to learn emergency procedures, countering a stall, putting the plane into a spin and getting out of it again. She drank in everything Caitlin had to teach her and was so avid to go solo she bought her own Travel Air for $5000 (which was five times the average annual income in California).

Since she couldn't fly her new plane without a licence, she had a photograph taken of herself on her favourite horse, jumping over the tail section, with the caption: 'I'm in the air over my new Travel Air.'

Pancho soloed on 6 September 1928. Ben Caitlin had his first successful woman student, and was mightily relieved to pass her out. Pancho was back in the air five minutes after her solo flight, carrying a childhood friend, Nelse Griffin, who was so excited by the experience that he decided to try a little wing-walking. Any other new pilot would have forbidden such a crazy stunt: Pancho told him to go ahead and while the young man climbed out through the wires and struts, she brought the plane in low across the field, chasing her shadow at fifty feet. One Sunday soon after this, she flew over St James church, where the Rev. Rankin had just delivered a valedictory sermon before setting off for New York. In the silence of the following prayer the parishioners heard the sound of an aeroplane engine getting louder and louder – and louder. Pancho circled the church tower three times: she knew she didn't have to be down there any more, she was in heaven right where she was, or as she put it in her own, inimitable way, she was as happy 'as a sex maniac in a whorehouse'.

Pancho realised that even with the Mystery Circus under her belt, she was still an inexperienced pilot, and set about getting a few long-distance flights in her logbook. She also

29

bought herself a newer, faster plane, a Travel Air Speedwing, which sounded like the kind of craft she'd enjoy. And she did, flying, throughout 1928–29, hundreds of miles up and down the coast of California, airfield hopping, following the coast road, dropping in on old friends and, more important, making new ones and establishing a reputation in the aviation community.

Her instructor, Ben Caitlin, had moved his base of operations and was working out of Carpentaria, where he was managing the local airfield. He was no longer so anti-women, at least, not anti-Pancho; she had won his respect with her tough, no-nonsense attitude and now he welcomed her into the back office where, after the day's business was over, he and the boys would sit around and talk flying. And she loved 'hangar flying' as it was known, with whichever pilots might be passing or stopping over for the night. They would talk about engines and wind speed, the problems of navigation, of keeping the craft in the air in stormy conditions, about the risks of night flying when a pilot might find himself following a set of car headlights while looking for somewhere to land. And always, at the back of everyone's mind, was the simple fact that they were engaged in one of the most dangerous occupations in the world and that each and every one of them found this terrifying and exhilarating in equal measure. The daily risk was the whole point – it meant, as someone once said after a party, that 'I guess we have more fun than the people.'

The shifting nightly gathering of pilots even became an informal but very exclusive club, the Short Snorts; the membership card was a dollar bill signed by each one of the group. Anyone who turned up without his bill was liable to find himself landed with the evening's bar bill – and that could be expensive! Pancho never forgot her banknote but, somehow still ended up paying, quite happily, for many evenings. She had inherited two mansions by this time, one at San Marino, which became an unofficial clubhouse for the Short Snorts (it also became a glorified bar where the liquor never ran out and rarely had to be paid for) and the other, a cliffside dwelling at Laguna Beach. Here she extended her Hollywood contacts and began hosting parties for her more respectable, but no less wild, friends. One night, a Mexican Air Force colonel ate the underwear of a visiting actress (apparently he was known as a serial pants eater, which may help explain the poor record of the Mexican air service) and evenings seldom ended without the swimming pool filled with tuxedoed matinee idols and starlets dressed by Adrien of Hollywood. Pancho's old pal Erich von Stroheim was an early caller and soon became a fixture, sporting his monocle, Prussian-style shaven neck and riding crop. One of his favourite pastimes was to stir up quarrels with his friends, thinking it 'sharpened them up' and gave them a proper respect for his aristocratic breeding, which was, appropriately in this movie-land setting, as fake as the 'von' in front of his name. At one party, he chose to tease Pancho about her Mexico trip, asking if she'd really posed as a man. She told him it was all true. He grinned and shrugged in his Viennese way and told her that it was impossible for any woman to convincingly ape a man. Pancho told him to try *this* and see what kind of man she made, and hit him hard enough to knock him on his bottom.

Other guests included the screen idol Ramon Navarro, a man in the flashing-teeth-buccaneer mould of Don Rockwell, and Pancho found herself more than a little attracted to him. Unfortunately, his style didn't include women, at least not between the sheets, so Pancho had to be content with his friendship; and since he was a pilot, that was no problem at all. Norma Shearer, and the young John Wayne, then making his first picture; famous aviators like Jimmy Doolittle and Roscoe Turner, who used to fly with his pet lion-cub as companion: they all called in and stayed over for the non-stop party – the young, the beautiful, the rich and famous.

Pancho was young and rich, but she wanted to up her average. She'd never be beautiful, though she dressed in high Hollywood style – when she wasn't in slacks and flying jacket – but she could become famous, and the way for a flyer to do that, in 1929, was to set records and win races. And right there, in the local paper, was a story about record-breaking pilot Bobbi Trout, who was issuing a challenge to other women pilots to compete in a forty-mile air race, back and forth between Van Nuys and Glendale. It would be, Trout said, the very first women's air race in history, and that sounded pretty good to Pancho. She signed up there and then. The only other taker was Margaret Parry, a local aviatrix and airport owner.

Pancho had no intention of losing the race and took no chances. Her Travel Air was far more powerful than either of the other planes; she won comfortably but, more profitably, the race marked the beginning of a long friendship with Bobbi Trout and gained Pancho a good deal of the publicity she craved. She was also beginning to earn a good living (not that she needed it with her inheritance coming through) working as a test pilot for Lockheed and Beechcraft. The big companies had realised the publicity value in employing female aviators as representatives, flying saleswomen, executives and test pilots and many young women found this a useful way to get their hands on planes of a quality and class far outside their normal experience. But nobody ever got famous by going to the office every day – even a flying office – and Pancho was still looking round for something she could really get her racing teeth into; like, for instance, a 3000 mile derby across the continental United States: the so-called Powder Puff Derby.

The entrance qualifications, a licence and at least one hundred hours flying time, presented no problems for Pancho, she'd racked up far more than that over the past few months. She was also confident that her Travel Air 400 with its 200 hp Wright Whirlwind engine would stand a good chance of being placed, maybe even of winning, although she had to admit, if only to herself, there were pilots with far more experience than she'd gained: Marvel Crossen, Phoebe Omlie, who taught William Faulkner to fly (and must be held at least partly responsible for his worst novel, *Pylon*) and Louise Thaden; all multiple record-holders. If Pancho was to stand a real chance, she needed to prepare. Flying long distances across the country without getting lost was still a seat-of-the-pants operation, depending on roadmaps, landmarks, roads, towns and railway lines. Every flyer Pancho knew had endless stories of following the wrong tracks or the right tracks in the wrong direction, sometimes for hundreds of miles. Up there, in just about every kind of bad weather that a land mass as big as the United States could throw at you, it was not easy to get and stay oriented. There was glaring sun, storms, mist and Pancho's one weakness, low cloud – something she would never come to terms with; she always worried that somewhere in the cloud mass there would be a lurking mountain big enough to clout even her out the sky.

She wanted to familiarise herself with the course by over-flying it in advance (not against the rules) so she could have speed, times and landmarks worked out before the start. She could afford the time and the fuel more than many of the other entrants and, with preparation and the power of her engine, she was certain that this was going to be the race that would bring her to national prominence.

She lined up with the other starters and, when the pistol shot was broadcast from Cleveland, began to taxi forward behind her friend Marvel Crossen, with whom she would be rooming overnight during the race.

The first day's run was a short hop to San Bernadino, and Pancho made it in under 28 minutes, which put her in the lead – though she knew that that was just a warm-up and the real flying would start on the next leg, over the desert to Phoenix. A mid-flight stop was scheduled at a little desert strip at Calexico, and many of the flyers were concerned about whether the soft surface of the runway would provide enough support, particularly for the heavier planes. The race authorities refused to change the route until an angry Pancho collected signatures from every flyer and presented a petition that read very much like a demand. The stopover was changed to Yuma.

The second day was brutal, in more ways than one. By the midday stop Pancho had lost her lead, following the wrong set of railway tracks despite all her efforts to avoid this. At the end of the day, a pall was cast over the arrival at Phoenix by the news of Marvel Crossen's death. After the death of her roommate, Pancho herself was to fly only one more leg of the Derby.

The following day the competitors flew on to Pecos. A ruptured fuel line cost Pancho her place among the leaders but she was still confident she could make up any loss as she sighted the airfield that evening, coming in to make a perfect landing that suddenly turned into a nightmare. Out of nowhere, something hit the undercarriage, throwing the Travel Air violently to one side; the upper and lower wings slammed into the ground, crumpling as the plane spun out of control and skidded to a halt. As Pancho climbed out, she saw a car driving rapidly away; an over-eager spectator had actually driven on to the runway for a better view and parked just in the blind spot that was caused by the large engine cowling of the plane. The driver was never identified; fortunately for him. Pancho was unhurt, but the plane was a write-off. She said, 'I have flown that plane for 200 hours. I have flown it from coast to coast and from one border of the country to the other. Never before had I damaged it and, of course, my first accident would have to come on an occasion of this kind. I circled the field before landing, but that confounded automobile must have stayed right under the blind spot . . . and I never saw it. My right lower wing hit it. The plane described an exaggerated ground loop and the left wing hit.'

'That *confounded* auto?' Either Pancho was on her best behaviour that afternoon or the local reporter from *The Wichita Eagle* wasn't up to taking down her epithets verbatim. Biographer and friend Grover Ted Tate, in conversations with Pancho, heard a more believable version: 'Some damned harebrained sonofabitch drove his truck smack into the side of my airplane and knocked it for a loop!' Either way, she was out of the race and, after arranging for her plane to be shipped back to the Travel Air factory, she hitched a ride to Cleveland where she was at the finish line to welcome Louise Thaden as she won the first ever Woman's Continental Derby.

Pancho wasn't downcast. At the Cleveland National Air Races, held over the week following the Derby, the biggest success had been a new superfast single-cockpit racer designed by Walter Beech of Travel Air and known as the Mystery Ship. With a top speed of 200 mph and a 'clean' design with radical low wings, the scarlet monoplane grabbed the attention of every flyer on the scene as it beat even the fast army racers; it not only flew so fast that 'it took three men to see it', it also looked like a dream. Numbers were strictly limited but Pancho heard a rumour that one was available for the price of $12 500. If she could get hold of the craft, the speed records held by Louise Thaden and Phoebe Omlie, or anyone else for that matter, would tumble before her. She would be the fastest woman on the planet.

She bought the plane and, being Pancho, began to show it off around the airfields. Howard Hughes, then a young film producer, was impressed. He'd recently finished making a flying epic about the Great War, *Hell's Angels*; unfortunately, during the prolonged post-production period sound had arrived and silent movies were no longer popular. Hughes had decided to put a soundtrack on his film but the plane engines he'd recorded somehow didn't have the impact he wanted once they were transferred to the screen. When he heard Pancho pulling out the throttle on the 425 hp Wright engine he knew he had found his ideal sound; he employed her to stunt her plane around a tethered balloon with recording equipment in the basket, and for a couple of days she had fun swooping down on the balloon, climbing past it and producing just about every aircraft noise Hughes was ever likely to need, short of crashing.

The film was a big success and Pancho's stock in the exclusive world of stunt flyers began to rise. Her public persona also rose when in August 1930 she beat Amelia Earhart's speed record with a time, over a measured course, of 196 mph. A year later she lost the record to Ruth Nichols but gained a cross-country speed record, flying from Los

Angeles to Sacramento in 2 hours and 13 minutes. This earned her no money – the flight was for charity – but gained her a sponsorship deal with Union Oil. She appeared in a series of magazine adverts: 'Mrs Barnes knew that for maximum power, uniformity and dependability, Union-ethyl gasoline has no equal. We are glad that it met her expectations!' She also got a cup, presented by the Governor, inscribed 'America's fastest woman flyer'.

Everything was looking good for Pancho and she started looking around for a new lover, and thought she'd found him in Duncan Renaldo, another man in the physical mould of Don Rockwell; dashing, fiery with a dazzling smile and a film career that was gathering momentum all the time. Renaldo had started as an extra but rapidly became a film star, in the style of Ramon Navarro, though his tastes inclined towards women rather than men. But not, unfortunately, towards Pancho. Despite her enthusiasm – she gave him the run of her mansions and drinks cabinets and was never in the habit of sparing the dollars – he resisted her charms, if not her largesse, as long as it lasted. He could see as well as anybody that things were looking a little unstable in the money markets; it began to be apparent that the endless party that was the 1920s, for the young and rich, was about to come to an end with a hangover and economic depression.

The Dobbins' wealth provided a cushion for a while and Pancho carried on with her usual lifestyle; she could hardly have lived in any other manner. Money had never meant anything to her when it was there in abundance, and she wasn't about to change her way of thinking just because the dollar was sinking. She even took a short flight into politics, standing as candidate for Supervisor for Los Angeles Third District. She was supported by Hollywood and her flying friends and put out an election address, calling herself Florence rather than Pancho Barnes, stating that, among many other virtues, she had an excellent understanding of women's issues, economics and children's welfare. But not presumably of truth, since little Billy, getting bigger every day, would hardly have recognised his mother if she'd formally introduced herself.

She stood as an Independent – neither Republicans nor Democrats would have touched her – and though she gave a good show, sky-writing her name in smoke above the city, she had neither policies nor anything else to offer the voters. The explanation for the adventure may lie in the practice of bringing in an independent candidate to split the other side's vote in a tight seat. The ex-deputy Governor of the state, Buron Fitts, who was standing for the district attorney's office, was something of a political fixer and he had been the one to encourage Pancho to try out her political wings. After the election, which Pancho lost and Fitts' party organisation won overall, the new DA paid Pancho to fly him and his advisers to Mexico. All in all, it was probably a foregone conclusion and Pancho got some fun and some subsidised flying out of it.

She also got a visit to Mexico City where she was stopped at the door of a notorious brothel one night while Fitts and his buddies strode in. 'No ladies', she was told. She could have told the doorman that she *was* no lady, but he wasn't in the mood to listen. Next night Pancho appeared again, in the uniform of a Mexican Air Force Colonel (perhaps borrowed from the lingerie eater?) and strode in, slugging the doorman as she passed. She hired a couple of girls and took them back to Fitts' hotel room but the new DA and his friends were too tired to perform, so Pancho bought the girls a drink and sent them back. All in all, Los Angeles Third District was probably lucky to escape getting Pancho as their Supervisor.

This was the last of the good times for a while. Even Pancho's wealth had its limits and, as stocks plunged, so her income began to dry up, and for the first time in her life she needed to earn money to eat. So she went back to the films with her plane and experience and joined up with a few of her old hangar buddies to form the Association of Motion Picture Pilots (AMPP). The union aimed to protect stunt flyers and their livelihood at a time when anyone with a plane would work for a few dollars a day and most producers were happy to hire them, despite their inexperience. The AMPP was able to bring some order to the stunt business, and save a lot of lives, since tyro pilots were cracking up or spinning in – the group never ever said 'crash' – at an alarming rate.

Not that the professionals were getting off scot-free: films about flying were popular, one of the few sure-fire money-makers Hollywood was producing, and the stunts demanded were increasingly foolhardy. Good flyers were getting injured or dying, and the AMPP decided to do something about it. They reasoned that the biggest disadvantage of getting killed was that you missed your own funeral, so they instituted a tradition of living funerals, where each member got to attend his own last party. As for making the stunts somehow less dangerous, they were pilots, the best of the best, the toughest of the tough and not one of them would ever turn his or her back on a stunt because it was too dangerous.

Apart from lending room for club meetings, which inevitably ended up as drunken parties, Pancho's most significant contribution to the group was her fight to force producers to pay a proper rate for stunt flyers on their films. Howard Hughes was shooting two films, back to back, using non-union labour. Pancho didn't have a lot of respect for the billionaire's penny-pinching: 'Howard Hughes is a two-for-a-nickel son of a bitch,' she said. Hughes returned the compliment and refused to back down on his pay levels. Pancho, for once in her life, realised that this was not the time to use her fists, satisfying as that might have been. It was a question of union organisation, and in 1932 the AMPP managed to push through a raft of minimum pay agreements and flying safeguards that became the industry norm.

Being part of the AMPP had given Pancho a taste for organisation and, as the 1930s progressed, she turned her attention to the military. After all, if her grandfather, Thaddeus Lowe, had practically invented the US Air Force, why shouldn't Pancho follow his footsteps and create the first real Women's Air Force? There had been a couple of earlier attempts to get women into some kind of auxiliary formation: the Betsy Ross Air Corps and the tautologically challenged Women's Aeronautical Air Force, both of which Pancho had joined and neither of which had produced anything of lasting value. So she decided to do something herself. She went to see an old friend, Army Air Force Colonel, later General, Hap Arnold and was introduced, by him, to LaVelle Sweeley, an aviatrix with experience, through her husband, of the Army Reserve. The two women got on well, found their ideas coincided and set about creating a Woman's Air Reserve. They were given support and facilities by the Army Air Force and soon began pulling in members from the west coast. A lot of film flyers joined; Vera Dawn Walker, Mary Iggens, Myrtle Mantz, Louise Thaden, Blanche Noyes and Bobbi Trout were also members; Amelia Earhart, though offering her support, predictably stayed clear of an organisation of which she would not be the public figurehead.

To be fair, Earhart may also have harboured certain doubts about Pancho as leader and flyer. She was in the middle of writing her book on flying and women pilots, *For The Love of It* and, while mentioning just about every prominent woman flyer in the country with approval, had not included Pancho at all. Always aware of publicity (married to a publicist, how could she not be?) she may not have wanted to criticise a fellow woman flyer, to avoid giving ammunition to the anti-women pilots lobby. She might also have wanted to preserve her reputation as the popular voice of women's flying, and a quarrel with Pancho Barnes, bound to be public and messy, would certainly have left her reputation muddy.

Pancho's response to Earhart's lack of enthusiasm – and she would certainly have read the book and noticed her exclusion – was uncharacteristically low-key at the time. Perhaps she too did not want to add fuel to the anti-women fire or, for once, reined in her indignation, knowing that Earhart was one of the most popular women in America. Privately, many years later, according to

Grover Ted Tate who knew her well, she considered Earhart 'a goddam robot', under the power of her manager and husband G. B. Putnam, who would 'wind her up and she'd go and do what he said. Whenever she fucked up, he would scold her like a child.' This may have been sour grapes or perhaps Pancho was seeing through the myth; she knew a bit about fakery herself and was always good at spotting it in others.

The man or woman who would hold the rope for a friend and hang on, despite knowing she was going to be pulled over the cliff edge too, was the kind of personality who gained her respect. Men like Doolittle and Yeager, women like Trout and Crossen – maybe, in the end, as she vanished over the Pacific, Earhart also joined that happy but suicidal band of brothers and sisters. And maybe not. Pancho always maintained, but never quite explained how, that she was present when a Navy radio operator picked up Earhart and Noonan's last signal: 'We heard her tell about being out of gas and heading for the sea', and Pancho never had any doubt that the plane went down over the ocean. Time has proved her right, though whether the story was true or just invented to emphasise the good judgement of a pilot who'd flown close to empty many times herself, we'll never, as with so many things in Pancho's life, quite be sure.

The Woman's Air Reserve was soon up and flying, with its own ranking system. Pancho, naturally was General Barnes. There was a training programme in place that included first aid and a mandate to push the National Air Licensing authorities into using the same guidelines and standards for men *and* women. The WAR's first mission was a three-plane cross-country flight from the west coast to Washington, sponsored by Gilmore Oil (in the Depression, fuel costs were a major burden on private pilots), where the pilots, Pancho, Bobbi Trout and Mary Charles, planned to hold talks on the subject of equal licensing. This, however, as far as General Barnes was concerned, was not the only or even the main purpose of the flight. Her old friend and hoped-for lover, actor Duncan Renaldo, had fallen foul of the immigration laws – it turned out he wasn't an American born of Scots–Spanish parents in California, but a Romanian who had entered the country illegally. He had just finished shooting the film *Trader Horn* on location in Africa and, on arriving back in the States, had been arrested, charged, tried and sentenced to deportation. He appealed against the judgment but had been turned down. Pancho knew her feelings for Renaldo were not and never would be reciprocated but he was still a friend, and she was never one to let a friend down. Deciding that the only recourse was a direct appeal to the President, she called in a few favours from old family friends and got in to see him. She put her case for Renaldo as forcefully as usual. Roosevelt promised to look into the matter. On the way home the pilots stopped over in New York to attend a civic function in their honour and faced being arrested for breaking a local ordinance forbidding women to appear in public in men's clothing. As Captain Bobbi Trout recalls, explanations were offered to the offended society matrons and, in the face of the patriotic trio, prejudice was overthrown and the women of the Air Reserve were allowed to attend in their full dress uniforms of horizon blue jackets and trousers, black Sam Brown belts, ties, berets and puttees.

Back in California, Pancho heard that the deportation order against Renaldo had been set aside, although he would have to serve time for the offence. Renaldo was later to achieve fame as the Cisco Kid, whose boast was that he defeated the baddies using intelligence rather than violence. When he was caught in a tight corner, it was often his partner, Pancho, who got him out of trouble – a tribute, perhaps, to General Barnes and her rescue mission.

By the mid-1930s, Pancho was beginning to feel the effects of the Depression; ready money was short, credit was non-existent and the property market had dropped through the floor. Her cliffside house had been reclaimed by the bank after she'd defaulted on the mortgage; she still had her San Marino mansion but didn't want to sell it if she could avoid it, though she couldn't really afford the upkeep. Stunt work was drying up – there were younger, sharper, more skilled pilots in the AMPP – and the public were no longer flocking to the air circuses. Pylon racing was still drawing the crowds but this had never been Pancho's thing; she preferred cross-country flying but, as the years passed, so the records fell and the distances increased. To get noticed, younger flyers, both male and female, were now setting out to cross the world and back. To cover everyday expenses, she borrowed $5000 from fellow flyer Paul Mantz, putting up her beloved Mystery Ship as collateral. He would have the use of it and would keep it in good flying condition, while she could use it any time she needed, within reason.

Pancho's solution to her financial problems was to fall back on her old skills of breeding and raising horses, but it was clear that coastal California was going to be too expensive for such an operation. She needed to relocate, and she knew just the place. Antelope Valley lies about 75 miles north of Los Angeles; it's a desolate place, part of the Mojave desert, without vegetation or water, the tiny population scattered over a large area, the main occupation alfalfa farming. Over the millennia, wind erosion and evaporation in the great heat had created a number of so-called 'dry lakes', where the salt surface was hard and flat enough to make almost perfect landing sites. These conditions had drawn the US Air Force to the area to set up a training programme at Muroc, later renamed Edwards Air Force Base. In the mid-1930s the base was no more than a collection of tents, a few huts and a fuel dump where pilots could fly in, refuel and practise their bombing runs over the desert but in years to come, the base would grow into one of the most important flight research centres in the USA. It would also prove both a blessing and a curse to Pancho.

She had first seen the area when she'd overflown it, and noticed how suitable it would be for an emergency landing. Now, looking to relocate, she found a ranch at Muroc, which was being offered in exchange for a city property. She flew up to look it over, landing on a perfect dry lake surface that was part of the property. The owner had drilled for and found abundant water – it flowed far underground – and created something of an oasis in the desert. Pancho liked what she saw and thought that, with a bit of work, she could turn the ranch into something that would yield a good living and, more important, be fun. The natural landing strip would mean friends could fly in when they wanted; all she would have to do was create a place they'd want to visit, and she'd never had any trouble doing that. She still owned an office building in LA and she offered it as her part of the exchange. The farmer accepted and she was now owner of Rancho Oro Verde, a cluster of buildings and a few hundred acres of the Mojave desert.

It didn't stay that way for long. Forced to rely on her own resources, without the cushion of wealth, Pancho reacted very much as she had in Mexico. She started having fun planning and then building a series of guest huts – and huts they were, though in later years they became increasingly more luxurious – and setting herself up as an alfalfa grower and dairy farmer. She began to supply the nearby Air Force base with milk, just as it began supplying guests for her bar and restaurant and swill for the hogs she bought in, and sold back, as meat, to the base. It was a perfect system, as long as peace existed between Pancho and the Air Force.

She had a new lover, Logan 'Granny' Nourse, a rangy, good-looking rancher who had plans of his own but enjoyed Pancho's company so much he was happy to put them on hold and help run Oro Verde. Something he didn't appreciate was the habit of the Air Force pilots of using his truck as a target for their practice bombing runs, dropping sacks of sand or even cement around him as he desperately tried to outrun them.

Pancho, predictably, loved the pilots and the wilder they were, the better she liked it; evenings at the ranch could last into the next day, when the pilots made their sorry way back to base – Edwards was still a tent town at this time – hungover and unsteady. Not that Pancho had any problems keeping order. At a local bar one night, not hers, some fellow started causing trouble and Pancho, after asking him politely to 'fuck off and let me eat my dinner in peace', and not getting an acceptable response, hit him so hard under the chin that she lifted him clean across the table without so much as spilling a drop of booze. The story was probably an exaggeration – but not much of one, and her reputation began to get around. She was tough and, she was doing ranch work, getting tougher every day. If she had any regrets, it was that she had less time for flying: but if she couldn't fly out, friends could fly in. She set up an airstrip with tying-down facilities and hangar/huts and offered the service as part of the overnight deal for visitors.

The Reverend Rankin Barnes had come to a decision of his own. He'd fallen in love and wanted to marry, despite the problems a divorce would cause to his career ambitions. Maybe he'd had enough of climbing the ecclesiastical greasy pole or perhaps the love of a good woman was worth the sacrifice. Either way, he asked Pancho for a divorce and she was happy to oblige. Maybe *she* was getting sentimental, but the idea of getting married to Duncan Renaldo, who'd just got out of jail, was beginning to look good; though not, unfortunately to the future Cisco Kid. The only desert he wanted to see was on a back lot in Hollywood; he remained friends with Pancho but never succumbed to her charm. Granny Nourse also had other things to do and, deciding that the ranch was now well established, he set off for home, leaving Pancho with her son Billy. Now aged 14, he came to stay and, in some accounts, enjoyed life at the ranch, in others, he hated it like rattlesnake poison.

Pancho loved the ranch and, as war in Europe began to seem increasingly likely, and the Air Force strengthened its presence at Edwards, she decided to expand her facilities still further. The huts were improved and became lodges, the bar and dining-room extended, a swimming pool, surrounded by trees, was built, stables were put up and riding horses bought: her brochures now called the place Rancho Pancho, the Flying Dude Ranch. She also found a new lover, Mac Mckendry, who moved in with his son; the boy received rather more of Pancho's care and attention than Pancho's own son, though considering her advice to the child's teacher ('If he doesn't behave, beat the shit outta him') Billy probably wasn't that upset about it.

During the war years Edwards grew into a major Air Force facility; business boomed and Pancho began to think it might boom even more if she got rid of her dairy and opened a casino. Setting the idea of a steady earner against the excitement of a fast buck was never any contest with her and the cows went. Unfortunately, the gamblers didn't arrive and after a few months the gambling den closed down. Pancho wasn't concerned; she'd always rather look to the future than look back at the past; and the future was about to arrive on her doorstep with a bang!

The sound barrier was considered by many in aviation to be an impassable wall through which man and plane would never burst. Not so by the pilots of the experimental flight programme at Edwards. The best of the best, with more 'right stuff' than most of them really knew what to do with, they brought, in the late 1940s and early 1950s, an entirely new atmosphere to the ranch and Pancho responded by creating an entirely new ranch.

Nobody quite knows how the Happy Bottom Riding Club really got its name but most people who were there agree it probably had more to do with the riding that was available than the beautiful hostesses who might or might not – depending on where you stood during various later court cases – also be available. Suffice to say that Pancho employed a number of young women to work as waitresses, all of whom shared the name Smith: January Smith, Tuesday Smith, Nevada Smith and so on. Pancho reckoned it added an air of mystery to the place. As for the boys who came for the girls, she reckoned that if they were old enough to risk their lives jockeying jets out in the wild blue yonder, they were old enough to handle some feminine company. The management put up a sign that said: WE ARE NOT RESPONSIBLE FOR THE HUSTLING AND BUSTLING THAT MAY GO ON HERE. LOTS OF

PEOPLE BUSTLE AND SOME HUSTLE, BUT THAT'S THEIR BUSINESS AND A VERY OLD ONE.

By the late 1940s and early 1950s, many of the test pilots had brought their families out to the housing set up by the Air Force and though, special events apart, they rarely took them to Pancho's, her account is almost certainly as true as anything she ever said. What the pilots did at the Riding Club, like most pilots in most bars the world over, was drink and talk about flying, and remember dead buddies from the war and from the X-1 and X-2 programmes.

The club was really a club within a club, for flyers only, named the Blow and Go. It had membership cards and the first was given to Jimmy Doolittle, a hero of the past, and the second to a hero of the future, one of the very few human beings Pancho ever admired unreservedly, Chuck Yeager. One night a civilian visitor to the club expressed some doubt about the ability of Yeager and his co-pilot Bob Hoover to fly the Bell Experimental Plane. Pancho responded: 'These two can fly right up your ass and tickle your right eyeball, and you wouldn't even know why you were farting shock waves.'

It was at Pancho's, on the night before he was to fly the Bell X-1 faster than sound for the first time, that Yeager had his famous accident. Cantering back after a desert trip, in the dark, he misjudged the gate of the corral and fell off his horse, breaking two ribs. He realised that the moment he reported the injury, he'd be out of the cockpit, so he didn't; he got himself tightly bandaged by a local doctor, went back to base and flew the next morning, using a sawn-off broom handle to close the cockpit catch. The X-1, more like a rocket with wings than a plane, was dropped from a converted bomber and Yeager blasted off into history, breaking the sound barrier with a boom that echoed across the salt flats of the Mojave desert.

Short of breaking the barrier herself, Yeager's flight was one of the great moments in Pancho's life and the celebrations that night were pretty great too. It was the climax of the Happy Bottom Riding Club and though things appeared to go on the same way for years to come, the edge of the envelope had been touched and what had gone up, was now beginning to come down.

The descent was hard to spot from the ground. The guests kept arriving and parties went on; Pancho moved into the rodeo business, which allowed her to indulge her old passion and expertise for horses. Leaflets were sent out advertising the attraction, and thousands attended the three-day events – one hell of a lot of fun was had by all and the ranch was admitted to the Rodeo Society of America. It was fun but, in the end, not profitable. Pancho had never really understood the concept of breaking even, never mind making money. She was happy as long as she could keep going, and keep on having fun and keep on helping out when it was needed. More than one of the test pilots found himself and his whole family being put up at Pancho's expense while house hunting in the vicinity, and it was a rare drifter or down-and-out who did not receive at least a handout and more likely a bed in the bunkhouse, a decent meal (and meals at the Club were decent enough to feed two to bursting) and the offer of a job if it looked like the fellow could put in an honest day. That was all Pancho ever really asked of anyone – honesty: in their flying, in their stories (more or less), and in their character. It was not all the Internal Revenue Service asked: they wanted their taxes and Pancho was in constant trouble, since she was never up to date in either national or local taxes. As for personal loans, well, what kind of person would ever ask for their money back? She never would. However, Paul Mantz did; he asked for the repayment of his $5000, Pancho didn't have it and her Mystery Ship became his. She couldn't complain but, somewhere deep inside, she couldn't really understand how one pilot could take another's ship for any reason.

Paying tax didn't bother her over much, she was always ready to give other people handouts, so she reckoned someone would come to her rescue, if necessary. But things were changing all around her. As the records were knocked down in the skies over the Mojave desert, the complexion of Edwards Air Force Base began to alter: the cowboys and all their virtues were becoming old-fashioned. It was a new kind of military, facing the threat of Communism, with new leaders who didn't see things Pancho's way. The Cold War was not fun: it was a deadly serious business, at least in their opinion, and one of the things they needed was more land for training men to fly the big bombers. Ranchers close to the base found themselves receiving compulsory purchase orders; some were glad to go, since making a living out of the desert was hard, unforgiving work; others had grown to love the place and its spare beauty, the way it bloomed suddenly, once in a decade perhaps, after a storm, coming alive in an explosion of colour.

None of this made any difference to the military: they needed the space, they would have the space. The only question lay in the valuation and, as their grasp reached out toward the Happy Bottom Riding Club, in the determination of the landowner not to give in. For Pancho, the letter advising her that the military required her land was a double blow. There was the derisory amount they were offering for the business loss she would suffer but worse, far worse, was what she saw as a betrayal by an organisation she had always loved and supported: the Air Force.

Throughout her life as a pilot, she had known and respected men like Chuck Yeager and General 'Hap' Arnold, Second World War Air Force chief, and James Doolittle, leader of a daring bombing raid over Tokyo in the early days of the war; they were her kind of guys. You could trust them because there was not one speck of bullshit in them. And now, here she was, facing a vast mountain of legal bullshit being shovelled over her ranch by the Air Force itself.

She still flew, though her case against the military, in which she represented herself, began to take up more and more time and money and keeping up maintenance on her planes became harder. In the early 1950s she took her old desert buddy Roger Chute down to Mexico where they flew around mountains and Pancho landed on a tiny island off Mazatlán, simply because nobody had been crazy enough to do it before. In typical Pancho fashion, though Chute didn't know it till near the end of the trip, she was flying without a licence; she'd just neglected to renew it one year, maybe because she was getting worried about her health – the high blood pressure which had killed her mother was starting to be a serious problem for her too – or because the form-filling, time-consuming bureaucracy of the whole business was getting on her nerves or just because she didn't. After all, who would mind? She was Pancho Barnes. Roger Chute agreed, she hadn't changed a bit from the girl he'd known 30 years before, he said, only her reactions *were* slowing down a little so perhaps she'd fly with a little more care.

The Air Force still wanted her land but rather than wait for them to act, she moved first and took them to court. She said, 'I never ran away from a fight in my life and I'm sure as shit not running from these peckerwoods.'

Hearings followed, in which two, three or four Air Force attorneys faced one woman across the courtroom; the men in uniform or sober suits, Pancho in checked shirt, jeans and cowboy boots. She was no lawyer but she was shrewd enough to duck and weave and keep on the move, and if she never looked like winning, she didn't look ready to lose either.

After a number of skirmishes, she managed to pin the Air Force down and make them issue an offer of $250 000, which she promptly rejected. She was almost certainly right to do so; the development she'd funded at the Oro Verde Ranch and the business it had generated deserved a better price. But it wasn't really the money that was the issue. She was desperate to put off the day she would have to pack up and move. In her early life, houses and their grounds had always been there, with everything she had wanted, from an exercise track for her horses to a swimming pool on the edge of a cliff overlooking the sea. At Oro Verde she had found a working ranch and made it into 'practically a small village', her village; her world where guests flew in, where beautiful bar girls never broke the law, where pilots sat around all night and told flying stories. It was hers and she was going to lose it but in the end even that didn't matter.

Losing wasn't the point, giving up was: she wouldn't hand anything over to the government without a fight – even after a disaster that would have floored anyone else.

In late 1953 the ranch caught fire; it was thought that a drifter she'd taken in had upset a stove while drinking. In the dry desert heat, the flames spread rapidly from the tackroom and stables to the visitor cabins and then to the main house. Fire-fighters arrived but ran out of water; the swimming pool had been emptied for cleaning and a storage cistern, which held thousands of gallons, was overlooked. All of Pancho's horses were killed when the stables went up: saddles, equipment, Pancho's possessions – paintings, books, mementoes - were all burnt: the damage was estimated at $300 000. The fire chief wasn't convinced by the drunken drifter story; he thought that the fire had started in more than one place and had been encouraged to burn. There were those who said it was a deranged ex-sergeant from Edwards Base, who'd been connected with a number of unexplained fires and, subsequently, arrested for skulking around the ranch. There were even those who said Pancho had set it herself to collect on the insurance, although it would have been utterly out of character for her to have risked the lives of her beloved horses; she'd sooner have burnt herself, and she wasn't about to do that: the place might be ashes and charred wood, but she would keep up the struggle.

And she won. It was crazy, and it shouldn't have happened in the second half of the twentieth century, but after three years of fighting her own case, appearing day after day in courts all over the state, humping vast piles of paper around – her strength standing her in good stead here – three years of appearing in front of juries arguing her case, not always well or coherently, but passionately and, more important, convincingly, the ordinary people listened and found for her and judgment was issued in her favour. The government must up their price and pay, with interest, $400 000, for the ranch at Oro Verde. Of course, she lost in the end. She had to move out.

Money had never meant anything to her, it didn't now. She paid four years' of back-taxes, a huge sum, met the various legal expenses she'd run up, and moved north, close to the snowy peaks of the Sierra Nevada mountains, buying a small ranch called Gypsy Springs on land even less promising and more remote than Oro Verde. She bought the local (20 miles away) café and the run-down gas station, she bought a plane and a cabin cruiser, she bought a large colour television, even though there was no electricity to run it. She wasn't spending money, she was just having some fun; the ranch was desolate, water was short, the house was almost a ruin, without electricity or proper flooring, but she'd been here before and built an oasis in the desert; she could do it again.

She bought horses and hogs, meaning to start up her recycling operations again, but here there was no Air Force Base to act as her customer, and few visitors were prepared to make the long trek out to the mountain landscape. Mortgage payments were missed, bills ignored, tax demands lost or chewed up by the dogs she kept. Her old friends were doing just that: getting older, and they now had families and career responsibilities. Only Pancho was the same, but she wasn't: in the late 1950s she got breast cancer and had a double mastectomy which, in typical fashion, she'd show to anyone, whether they wanted to see or not. The doctors told her that due to muscle damage sustained during the operation on her right breast, she'd lose the strength in her right arm. She told them: no way. She exercised the arm constantly, ignoring the pain, until she got full use of it back; she could give up anything – her house, flying, breasts but she would not give up her own physical strength. The only thing that could take it away from her was time and illness, and she struggled with both of these during the early 1960s; there was also an acrimonious divorce with Mac Mckendrick, whom she had married along the way.

She didn't give up, but she did start to look in the other direction more and more, ignoring what she couldn't avoid as goods and property were sold to pay tax and bills. She just wasn't feeling the way she used to, which wasn't surprising: she had thyroid trouble and she was close to dying, unable, as the days went by, to do much more than drag herself out of bed in the morning to feed her dogs, then drag herself back without eating. And she would have died had not Ted Tate, a young aeroplane engineer who had known of Pancho since the Happy Bottom days, called in on the off-chance of seeing her. There was a lot of yelling back and forth through the door of the shack – for some reason Pancho just wouldn't come out until finally she asked the young man if he was visiting formally or informally. Looking around him, at the scrubland, at the tiny, almost ruined shack, the dogs roaming, the ancient truck, he asked what on earth difference it made. Pancho, with the hauteur of a queen, informed him that since her recent operation, she only put her rubber tits on for formal visiting – and if Tate didn't mind, she'd leave them off today.

She eventually came out, and Tate got her to hospital where treatment allowed her remarkable constitution to gather its strength and fight back. When she got out, Tate told her that old buddies from the experimental programme at Edwards Air Force Base had instituted a Pancho Barnes First Citizen of Edwards celebration in her honour. The bar was open, there were friends to drink with and, most important of all, the chance to sit around and indulge in a bit of hangar flying and tell stories of the Happy Bottom Riding Club and the night when two businessmen ordered up the best dishes the house could provide and got two of the hostesses served in vast bread rolls; or the morning Yeager buzzed the ranch house so closely that tiles bounced off the roof and the base commander, who should never have been there at all, complained that his 'sleep' had been disturbed.

It was a great evening and Pancho, once again the centre of attention, loved every second. Then she went back to the desert.

In a way, things settled down. She began to accept her limitations. Her old Travel Air Mystery Ship, which had passed into the ownership of pilot Paul Mantz, was auctioned. Her son Billy, now a married airline professional and more or less reconciled with his mother, took her to the auction. There were a number of interested bidders (the ship was one of only two surviving) but Pancho's presence ensured that Billy's bid was successful: the Mystery Ship was back in the family.

Pancho wasn't really up to flying the plane any more. She took lessons for more than a year but the new Federal Avaiation Authority regulations proved a little too restrictive for her free spirit and, truth to tell, her reactions weren't up to it any more either. Roger Chute wrote to her, asking her if she really wanted to end such an illustrious career mashed up in a pile of wreckage.

After she had accepted this, and Chute was among the very few she respected enough to listen to, she returned to the desert and her dogs. She started an autobiography but she had never been the sort of character to stick with a project for that long; she lectured occasionally but mostly kept herself busy with the law, throwing herself with all her old energy into litigation against anyone who raised her ire, and that included a lot of people. Lesser matters, like housekeeping or clearing up behind herself, had never been particularly important in her life and that didn't change now: as she told a friend, there had always been someone there to do those jobs and if there wasn't any more, well, who cared?

In fact, quite a lot of people, who would otherwise have spent time with her, found her increasing eccentricity hard to take. Some close friends overlooked her habit of asking her driver to stop so she could get out of the car and squat to urinate at the side of the road, or her tendency to pull up her shirt so visitors could inspect her mastectomy scars or squeak her rubber tits; most found it impossible. When visiting or giving her talks – always without a script, just pulling great story after great story out of her memory – she would put on a dress and a wig and at least pass for respectable, but at home, in her little stone shack, she let things go because she really didn't care. Just as she'd never been the kind of teenage Pasadena debutante her family wanted, so she simply couldn't be the kind of elderly heroine of the skies her admirers expected.

She died alone at the age of 70. She was due to give a talk to a local historical society and, when she didn't turn up, her son Billy went to visit her and found her. Permission was granted by the Air Force for a fly-by over the site of the Happy Bottom Riding Club, where her ashes were scattered; and a memorial service was held at which Jimmy Doolittle spoke the eulogy. He ended, 'I can just see her up there at this very minute. In her inimitable way, with a wry smile . . . (watching us now) remarking, "I wondered what the little bald-headed old bastard was going to say".'

She was who she was – difficult, sometimes impossible, selfish and generous to a fault, a tough enemy but a friend who would be loyal to the end, a bad mother, a terrible wife, but above all, a pilot and some kind of a woman.

3
Bicycle dreams

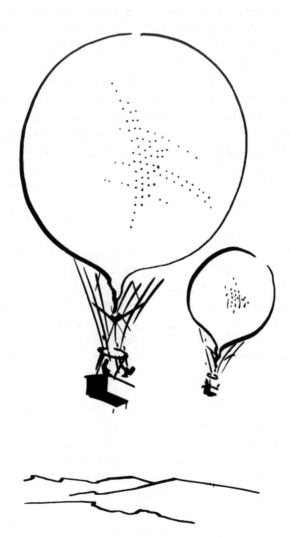

S o how did it all begin?
Perhaps . . . with hot air . . .
Joseph and Etienne Montgolfier were, so the story goes, waiting around the laundry room one day for someone to wash a shirt (not them, of course) when a draught of hot air from the fire under the copper clothes boiler caught the garment – was it Joseph's, was it Etienne's? – filled it, and sent it floating up to the ceiling and sent the brothers rushing out to plan and make the world's first hot air balloon. So the story goes and, in the history of flight, there are a lot of stories (pilots call it hangar flying), some of which are true, some of which are good, and a few of which are both. But probably not this one.

Let's just say that natural curiosity caused the brothers, paper merchants from Annonay in France, to start exploring the effects of heat on what they called 'envelopes'. In the course of time, the brothers came to the conclusion that if they could enclose enough hot air, or gas as they conceived it, in a large and light enough container – and what better than paper and silk? – then they might get it to float. They tested their first balloon on 4 June 1783 in the marketplace at Avignon. The unmanned, paper and cloth envelope was tethered over a fire of wool and straw which would not, it was thought, create too violent a gas. When the guy ropes were released, the montgolfier, as it would become known, floated away, rising to about 6000 feet.

A few weeks after the Montgolfiers' launch, the physicist Professor August Charles ascended in a closed hydrogen balloon in the company of his assistant, Monsieur Robert. They reached almost 2000 feet, then descended into a Paris square. When M. Robert jumped out, the lightened balloon shot up again, taking Professor Charles with it. 'I passed in moments from summer to winter,' he wrote, 'and though I was cold, life was not insupportable in the basket.' Calculating from the barometer on board, the professor realised he had ascended to 10 000 feet and could see 'that I was the only person in sunlight, all the rest of the world being in the shade.' He was found safely a few hours after dark that night and was renowned for being the first man ever to see two sunsets on one day. Benjamin Franklin was part of the excited crowd who observed the ascent and when asked, by a blasé spectator, 'What good is it?' responded, 'what good is a baby?'

The Professor was fortunate he hadn't landed beyond the confines of Paris. An earlier, unmanned hydrogen balloon had drifted over the countryside and although Professor Charles had issued a flyer, warning people this strange object they might glimpse overhead, 'like a globe resembling the moon in eclipse', was not a portent but merely a machine of rubberised silk, the locals in the small village of Genoesse had either missed the notice or been unable to read it. As the huge round shadow raced across field, hedgerow and cobbled square, total panic broke out. This 'thing' was beyond anyone's experience. They simply could not comprehend it. So, in the tradition of the European peasant, they grabbed pitchforks, flaming torches and the local priest and set off in pursuit, crying vengeance on the unnatural horror.

After a long chase, the balloon began to lose buoyancy and came within their enraged reach. Pitchforks, stones and burning brands were thrust at it to little avail, as the strong fabric and the lightness of the balloon allowed it to retreat upwards at every buffet, until a local hunter drew a bead on it with his musket and shot it dead. It expired with a ghastly wail and a foul odour. As it collapsed, the village rushed forward and terrible scenes of cloth-slashing and stamping ensued.

Just as, a few years later, the notion of high-speed rail travel induced thoughts of bodily injury – humankind was never meant to move beyond the pace of a fast horse – so the idea of ascending into the empyrean realm caused equal unease in general public and experts alike. Fortunately, neither public nor expert opinion has ever been enough to put the adventurous off trying something new – especially if it seemed really foolhardy.

And so . . . on what is recorded as a fine day in the summer of 1784, Madame Elisabeth Thible clambered into the commodious basket of a balloon under the direction of Monsieur Fleurant. It was an event – the King of Sweden was present, so perhaps not a huge event – and Mme Thible displayed commendable fortitude as the ropes were cast off, the brazier stoked and the craft began to rise free and untrammelled into the warm June air. She was dressed to the nines and hanging on to her hat as the montgolfier ascended to an altitude of 8500 feet and Mme Thible began to sing. We don't know what she sang and neither, given the height of her performance, did the King of Sweden but while the monarch has been forgotten, despite lending his name, *Gustave*, to the balloon, the lady is remembered as the first woman to make an untethered flight of any kind.

A month before this, three of her friends had sat in the car of a tethered montgolfier floating over the Faubourg St-Antonie and thought themselves very daring indeed. Mme Thible's achievement was of quite a different order: at the mercy of wind and weather, landing who knows where – only the year before, the very first aeronauts (sent up by the Montgolfiers) had, on landing, been tipped out into the upper branches of an oak tree. The sheep was bruised but had no dignity to lose, the cock fluttered like fury and landed more or less safely, and the duck leads us to ask why it sat there in the basket for the flight without leaving, as any sensible bird might.

Mme Thible never went up again, but once was enough to set a fashion and soon three or four others were thinking of trying their luck in the chariots of rival balloonists: a 15-year-old French girl ascended over England with Jean-Pierre Blanchard but regretted it at about 30 feet and started screaming. Mrs Letitia Anne Sage was made of sterner stuff. Like many an actress before and since, when asked if she could provide the balloonist Lunardi with beauty and ballast, said, of course she could. And she did, ascending in the summer of 1785 with two male companions; in fact they didn't ascend very far – there was too much weight in the basket and although Mrs Sage, a handsome fine-figured woman, was providing most of it, Lunardi the aeronaut established a principle that was to underline much of the early history of aviation: forget the science, go for the publicity. He and one of the male passengers got out, leaving Mrs Sage and a certain Mr Biggin, a small man, to fly on their own. And away they floated, off into the blue, making an exit that any actress in the land would have envied.

The event was recorded by the painter Rigaud, who shows the travellers reclining on a banquette in a spacious and thoroughly respectable (for an actress – no English *lady* would have done it) chariot, railed and curtained, so the gaze of the *canaille* need not bother the travellers. Mrs Sage is seated in the centre, smiling benignly, waving a plump hand as she passes over. To her right, Mr Biggin has a rather queasy expression on his thin face, while to her left, splendid in uniform and flourishing, as it were, his credentials, stands Lunardi, who wasn't there at all but since he needed the publicity and commissioned the picture, had himself painted in for the occasion. It did him no harm at all in his rivalry with J-P Blanchard (passengers to date: one frightened teenager and a 'small, ugly, nervous wife') to have not only the picture to display but an account from the fair hand of Letitia Sage herself, opining that aloft, she had been 'better pleased than ever I was with any former event in my life.' The schoolboys of Harrow were pretty pleased too when their daily routine of beatings, Latin and more beatings was interrupted by the arrival of a goddess in a machine. They chaired her into the town and gave her a slap-up meal. Her career improved no end after this daring expedition, though she never made another ascent.

The first female aeronaut to control her own vehicle turned out to be the small, ugly, nervous Madame Sophie Blanchard. Her husband, Jean-Pierre, had set a fine example of Anglo-French *concord* by making, in 1875, the first crossing of the English Channel in the company of an English passenger, Doctor Jeffries who, after a stomach-sinking, wave-touching scare put on his cork jacket, threw his coat and trousers overboard and felt as merry and untroubled as a finch. After her husband's death from a heart attack – first coronary in a balloon: 1908 – Mme Blanchard took up his mission. By all reports, a woman who found life at ground level hard to cope with, she experienced a sort of apotheosis in the silence and space of the sky (as many have since) and came to feel it was her proper sphere of being. She became celebrated as Europe's first professional aeronaut and was recognised as such by the Emperor Napoleon and the people of Paris who, according to a contemporary report, loved both her beauty and her daring. So perhaps the 'small, ugly and nervous' comes from the hand of a rival or an Englishman; but then the revolutionary balloonist Citizen Henry was always described as 'the young and beautiful', so perhaps contemporary reporters simply could not bear the idea of a French woman being elevated who was not young and beautiful?

Whatever her looks and character, Mme Blanchard had a wide-ranging intelligence and was fascinated by all aspects of the airy realm. During the early years of the nineteenth century she began conducting experiments with parachutes, sending a puppy down to earth on a number of occasions – puppy always arriving unharmed, of course. In 1819, she ascended from the Tivoli Gardens in Paris to give a night-time fireworks display in a hydrogen-filled balloon. 'Inflammable air' had been discovered by Henry Cavendish in the 1760s; it was, as Professor 'Two Sunsets' Charles always maintained, far more convenient for aeronauts than tending a fire but it was also highly inflammable. Before the start of the display, the balloon was hidden for a moment by low cloud, then, 'it reappeared, to the horror of the spectators, in a ball of flames. There was a terrible pause, then Mme Blanchard, caught up in the netting of her balloon, fell with a crash upon the slanting roof of a house in the Rue de Provence and then into the street, where she was taken up a shattered corpse'; thus achieving the distinction of first woman to die in a balloon.

During the nineteenth century, as the balloons went up, so the parachutes began to come down. The first jumps were performed in 1797 by André Garnerin, who also persuaded the 'young and beautiful aerial nymph' Citizen Henry (again) to make a radical jump – out of a balloon basket and into marriage. The marriage lasted, the jumping stopped, parachutes were still in their infancy and the death toll among experimenters was almost as high as it was among revolutionaries. It wasn't until one hundred years later that jumping became commonplace, at least as an entertainment – albeit one that still produced fatalities – a female jumper fell to her death in Peterborough in 1895 attempting a two-person descent with a male companion who, selfishly, survived.

The most celebrated of the Edwardian parachutists was Dolly Shepherd, a teenage waitress at the Alexandra Palace, north London's pleasure and leisure centre. Dolly was serving tea when she overheard a conversation between the Palace manager and Colonel Samuel Franklyn Cody, a Buffalo Bill lookalike who ran his own Wild West show and was also a brilliant designer of kites and planes. It was said he was as important to the development of British aviation as the Wright brothers were to America, and he made the first powered flight in Britain in 1908 at Farnborough. Designs from his workshop, including a monoplane and a sea-plane, were seminal in the formation of the Royal Flying Corps and the Royal Navy Air Service and without his work Great Britain would have been even less prepared than it was for the air war of 1914. Cody's biplane, the Flying Cathedral, was the largest then in existence and, on his death after a crash in 1913, 50 000 people attended his funeral and the King sent a personal message of condolence to his widow Lela, the first woman to make a non-balloon flight, in one of her husband's observation kites in 1902.

An odd mixture of brilliant engineer, showman and Elvis impersonator, the Colonel wore the whole shebang: fringed buckskins, flowing blond locks, spurred boots and a pair of gleaming six-shooters, with which he performed great feats of marksmanship, shooting eggs off Lela's head – sometimes missing, not by much but by enough to cause a crease and a headache and an extreme and understandable reluctance to stand in front of Sam's pistols again. Ever.

He was complaining about this over tea and crumpets when, there and then, young Dolly volunteered to take Mrs Cody's place. It was one of those spontaneous moments, and when the Colonel revealed that he always did his target shooting blindfold, Dolly must have wanted to kick herself seven ways till Sunday. She didn't – stalwart and true to her word, and perhaps suspecting that this was a way out of her boring suburban life as a waitress (one way or another), she stood firm on the boards that night and, while at least one member of the audience fainted, and the Colonel aimed and squinted under the bandanna, got egg all over her face.

The Colonel was grateful – in time he persuaded Mrs Cody to get back into the saddle – and as a reward for her help gave Dolly a tour of his workshop. The theatre manager, M. Gaudron, accompanied the pair: Dolly was an attractive and vivacious young woman, as photographs testify. She also asked intelligent questions and, at the end of the tour, noting her interest and her looks, Gaudron asked if she'd like to try aerialism herself. The Palace was noted for weekend displays of balloon jumping, which drew large crowds. Dolly wasn't sure exactly what he was offering, but intuited that it was definitely going to be better than being omletted nightly, and she accepted a position as *ingénue* parachutist.

A star was born. At least, a star was trained in a remarkably short time: about half an hour, as Dolly records in her memoirs *When The 'Chute Went Up.* After learning how to fall, testing her grip, and introducing her to the jumper's uniform, a sort of blue boiler-suit with gold trim, she was taken off by Gaudron for her first jump.

The balloon went up to 2000 feet and, as soon as Gaudron saw a good field, Dolly came down. It's difficult to imagine what her feelings must have been – she writes about a mixture of fear and sheer exhilaration – but there were other opinions. Her aunt, with whom she lodged in Lewisham, threatened to cut her off entirely for doing something so . . . impossible. She relented, but only on condition that her niece never, ever mentioned parachutes in her hearing or the hearing of her friends or neighbours or even of any of the local tradesmen.

She might have kept quiet about it locally but for her thousands of fans, Dolly *was* a star – and, when necessary, she had a cool head too. On one flight, when the ripcord malfunctioned and her chute wouldn't separate (the balloon deflated automatically when the parachute detached) she found herself rising to 15 000 feet, freezing cold, almost losing consciousness as the air darkened around her. She hung on, singing hymns aloud until, hours later, the balloon finally descended. On another occasion, she was performing a double balloon jump with a new recruit to the troupe, Miss Louie May, and this time it was Louie's chute release that wouldn't function (you might have thought they would have double-checked). Louie was on her first ascent, the balloons were three times higher than they should have been and Dolly knew that her companion was going to let go any moment and fall to her death. She managed to jump across to Louie's balloon, get her out of the defective chute and, telling her to hang on tight, launched them both into space. The single parachute wasn't designed to take the weight of two and the landing was a terrible jolt, injuring Dolly's spine. Louie was fine. Although Dolly thought she might not walk again, let alone jump, she was back in the air within weeks.

She continued to delight her public until 1912 when a premonition told her that her next jump would be her last. She never made another jump and lived to the age of 85, leaving one to think that flyers are like pianists – they go early or they go late but rarely do they go in between.

Balloons float – they are lighter than air so they go up, they drift with the wind and they come down and in no real sense do they fly, any more than a cork sails. So where did it really begin?

It began with Sir George Caley and his gliders – or perhaps his principles of mechanical flight – since they laid down the fundamental facts: heavier than air flight could be achieved 'by making a surface support a given weight by the application of power to counter the resistance of air.' It was to explore these rules that he constructed, between 1790 and 1857, a whole series of gliders, some fanciful, with flapping wings, but most intensely practical, if somewhat terrifying to the pilots he persuaded to fly them. He started with models of gliders and helicopter-like designs, which prefigured the theory of propellers. In 1849 he built a full-size glider and launched it with a half-size human, a ten-year-old boy, aboard. Less than satisfied with the result, he built a stronger craft with an adjustable tail-plane and sent this one off with his coachman, John Appleby, as ballast; it frightened Appleby half to death and prompted him to quit the moment he landed, but it flew for a short distance down a hillside before crashing and achieved the first unpowered heavier than air flight.

Caley's writings were not widely published in his lifetime but, as the century progressed, his ideas began to percolate into the general consciousness of flying theorists across the continents. His fascination with aerodynamics and the potential for powered flight were particularly influential in stating not only the problem but advancing the idea that there could and would be a solution to it. He wrote, 'the noble art [of flight] will soon be within man's competence and we will eventually be able to travel with our families and baggage more safely than by sea, and at a speed of 20 to 100 miles an hour. All that is needed is an engine that can produce more power per minute than can the human muscular system.'

Where it almost began: everyone expected it to happen on 7 October 1903, on Washington's Potomac River, when the Great Aerodrome, the flying machine created by America's best-known and most respected aeronautical pioneer, Samuel Langley, was readied for takeoff. Secretary of the Smithsonian Institute, Langley was an immensely gifted and courageous man; for a well-known academic to risk his reputation in the business of heavier than air flight was seen as eccentric, if not downright bizarre. He devoted himself to creating a series of highly successful model gliders, which he called aerodromes, and then a second series of models (the largest with a wingspan of over 12 feet and weighing 30 pounds) powered by small steam-driven propellers. In 1896, in a test on the Potomac, watched by Alexander Graham Bell, a model aerodrome was launched by catapult from the top of a houseboat and succeeded in flying a mile and a half at a speed of 30 miles an hour.

No wonder the Wright Brothers were worried that October day when a pilot climbed aboard the Great Aerodrome and the catapult was wound back and ratcheted into position. Langley's craft weighed in at about the same as the Wrights' Flyer but its engine was both lighter and more powerful. Unfortunately, Langley himself was a theorist, brilliant but not practical: his models and the Great Aerodrome were built by assistants, he never had any plans to sit in his own ship and fly it. The construction was simply not up to par and the catapult, while fine for launching models, was totally unsuitable for a full-size machine. As soon as it was released, the Aerodrome flipped over the edge of the houseboat and into the water. A local reporter compared Langley's dream to a duck, which expired, sinking without even a valedictory quack into the icy Potomac. Nine days later, on 17 December, at 10.35 a.m., the two bicycle mechanics from Dayton, Ohio made the first of four manned flights.

The Wright Flyer was mounted on a track running along the beach near Kill Devil Hills in Carolina. It was a perfect spot for flight experimentation and the brothers had spent many months searching before finding it and setting up their summer operation. Over the winter they joined their sister Katherine, who ran the family bicycle business in Dayton. Orville was later to say, 'if ever the world thinks of us in connection with aviation, they must remember our sister Katherine', who not only contributed ideas to her brothers' work but put up a great deal of the finance that kept them going. Back in the small Ohio town, they ran their business and designed and built, in the upstairs workshops, many of the light, tough components that would go into the construction of a whole series of gliders and then, finally, their powered Flyers. They had learnt, from building bikes, the importance of the ratio of weight to strength to power – after all, no one wants to pedal a lump of heavy metal up a hill or find a wheel rim crumpling the first time they go over a bump in a track. They also learnt about safety (they would be flying their own craft) and, from the world around them and the research of earlier aeronautical pioneers, they gathered every speck of useful data they could find, leaving as little as possible to chance.

Phlegmatic men, both of them – when a young visitor to the workshop sat on Wilbur's favourite Stetson and squashed it flat, the incident passed over without a word – they were also intensely private, and later in life Orville was to insist that potential biographers concentrate not on the men but the machines. Wilbur was under no illusions about the risks they were taking. Introducing himself to flight pioneer and theorist Octave Chanute in 1901, he wrote of being *afflicted* with the belief that powered flight could be achieved and went on to say that his disease was well nigh incurable and would probably cost him not only his livelihood but also his life.

That December morning the brothers shook hands for longer than usual, a witness noted, as if they didn't want to let go of each other. Then Orville – they had flipped a coin for a previous failed attempt three days before and now the turn was his – climbed aboard the craft and tested the all-important controls for elevation and rudder. Wilbur encouraged the watchers, mostly local lifeboatmen (one of whom, a total amateur, had been chosen to work the still camera) to raise a cheer. The engine was started, and the Flyer began to move along the track. After about 40 feet it rose off the rails and flew for 120 feet before a maladjustment to one of the stabilisers brought it swooping down on to the sand, where a skid was cracked in landing. That was it – human beings had stepped into another element and after those 12 seconds nothing was ever going to be quite the same again.

The brothers sent a telegram home: 'Success four flights Thursday morning all against 21 mile wind started from level with engine power alone average speed through air 31 miles longest 57 seconds inform press home Christmas.' There's considerable confusion about just who did send the news to the press but the local *Virginian Pilot* published the scoop next morning, the 18th: 'FLYING MACHINE SOARS THREE MILES IN TEETH OF HIGH WIND OVER SAND HILLS AND WAVES AT KITTY HAWK ON CAROLINA COAST.' It is notable, in the following story, amid the hyperbole about shooting out into space and elevating propellers, that the press was back to its old habit of lauding the physical attributes of these new gods – no goddesses yet – of the air. Though this was an age when a respectable man would rarely be seen outdoors without his hat, balding Wilbur was apportioned 'straight, raven hued locks' while Orville became a figure of 'magnificent physique'. Wilbur was also given, apparently, to shouting 'Eureka', at suitable moments. It was to be the beginning of a long and often fraught relationship between aviation and the mass media. But not yet. After that first inflated and largely invented report, a silence fell. Other pioneers sent messages congratulating the brothers, and a few journals noted the event – the local Dayton paper, so an apocryphal tale has it, merely stated, 'Popular local bicycle merchants home for holidays.'

The brothers prepared and issued a detailed report on the flights in January 1904 which they sent to Associated Press with little response. In May they returned to the Carolina beach and prepared Flyer No. 2, to demonstrate to an invited audience, once and for all, the plain facts of flight. Conditions were against them, the craft didn't take off and there were no spectacular photographs and no serious scientific reports. Interest began to die and the brothers retired to their workshop where they began to refine their designs, with a view to producing a model which they could sell internationally and recoup some of the money they had spent.

Secretive as they were, reluctant to show detailed plans or to provide demonstrations, sales were slow. However, their achievement had been noted in Europe, particularly France, where the Aero Club de France was already offering prizes for distance and endurance flights. By the end of 1907, even America had begun to notice; the Federal government was asking manufacturers to put in tenders for the creation of a heavier than air machine. The Wrights realised that if they wanted to be part of this movement, they would have to return to Kill Devil Hills with their latest model and prove their claims once and for all. Which is exactly what they did, making a number of flights throughout May 1908, putting their craft through a whole series of tests as they soared 'every bit as gracefully as an eagle' above the watchers, some of whom were invited, and many, from the national press, who were observing from concealed positions. That did it, 'once and for all'.

The first woman to design a plane was a New York stenographer, E. Lillian Todd, who presented a scale model of her machine in 1906. She had overcome the problem of engine weight by an ingenious system that would allow the plane to power itself by means of its own momentum. In just about every account of Miss Todd's work, this is the point where the word 'unfortunately' pops up, always followed by: it didn't fly. However, not all her ideas were so impractical. She presented a plane to the Signal Corps of New York State, which became the first in the nation to have an air arm; and suggested the foldaway plane or glider, which would make for easier transport, an idea which was soon to find its day in the automotive 1920s. In 1908 she became the founder of the Junior Aero Club of America through which she was able to share her undoubted love of flying with thousands of young people, particularly girls, whom she was always ready to encourage into making a career out of flying – if not designing planes!

And those early women flyers were, so often, actresses or circus performers or motorists or the ones who wore the pants, or aristocrats; and almost without exception they were single-minded in pursuit of what they wanted. If they were practical as well, they had a good chance of living a long life and, if they were dreamers, they didn't, but in either case they all experienced something that could never be taken away from them: 'the rapture, the glory and the glamour of the very beginning'. The words are those of Gertrude Bacon, an English balloonist, researcher and aviatrix who, in the years before the First World War, published a flood of articles and photographs encouraging her countrymen – though not, to any great extent, her countrywomen – to take to the air. She felt that the British were backward in accepting the challenge the Wright brothers had set (Col. Cody, the most important of the 'British' pioneers, was an American) and pointed across the Channel, where the French were making all the running.

Elise Daroche was an actress; not a great actress, but her career was satisfactory and she knew all the personalities in the theatre world and lots of those in the arts. She'd painted and sculpted, she'd driven cars in competitions and escaped the Zola-esque fate of many a second-rank, good-looking actress, that of becoming a rich man's mistress. She demanded rather more of life, and when Wilbur Wright came to France to demonstrate the Wright Flyer, she was, like everyone else in the world of the arts and sciences, a fascinated observer. She'd recently changed her name (no unusual act in the theatre); it helped distance her from an upbringing as a plumber's daughter and gave her the foundation on which to build whatever legends she might choose to erect. She became Raymonde de Laroche and because she could cut a dash and wear clothes brilliantly, she also became a personality in the small, incestuous world of artistic Paris. And because she was a personality who appeared in all the right places, she got to take a free flight and she was enraptured by the experience. She said, 'This is the way everyone should travel.'

Shortly after this, she met the pioneer aeronaut and engineer Charles Voisin, who was as enraptured by her beauty as she was by the idea of flying; it wasn't much of a surprise, then, when he offered to teach her the necessary skills. Days later she arrived at the flying school he ran with his brother Gabriel and found herself sitting in a Voisin-Farman biplane. One of the first commercially available powered flyers in the world, the craft provided a good level seat, though only one, so Raymonde was forced to practise her taxiing up and down the training field while listening to shouted instructions. She did well and appeared to have a real feeling for the craft – which was fortunate since her plans for her first lesson didn't include getting out of the plane without having actually flown it. Despite *cries de horreur* from the instructor and ground staff, she opened up the throttle, turned into the wind and took off. An observer noted that 'the plane skimmed through the air for a few hundred yards and then settled down again.' By now everyone on the ground was cheering.

Over the following weeks she began to extend her flying range, practising take-offs and landings and despite a nasty spill which resulted in a broken collar-bone, she achieved her licence, the first to be awarded to a woman anywhere, on 8 March 1910. She'd been worried that during her convalescence one of her rivals would have taken the honour and it was a close thing – within the year, five more Frenchwomen had gained their licences. The press was fascinated by *la femme-oiseau*, as they immediately christened her. And it's interesting to note the worldwide press obsession with flinging *soubriquets* at any woman who left the ground: the Girl Hawk, the Flying Schoolgirl, the Cardinal of the Sky (red skirt and blouse and some mixed-up theology), the Tomboy of the Air, the Flying Flapper, the Dresden China Aviatrix (they were obviously running out of names by now), the Flying Duchess (she was a duchess and she flew!); the list goes on. Apart from the raven locked Orville Wright at Kitty Hawk and Jim Mollison, the Flying Scot, the chaps, on the whole, escaped the practice.

In her interviews Raymonde made light of the risks pilots faced. She told the hacks that controlling an aircraft was more a question of feel than of sheer physical strength and she wasn't taking any more risks than anyone else, she was just packing them into a shorter time span. Besides, she added, no one can avoid their fate: what will happen, will happen. The press loved it and awarded her the title of Baroness there and then. She was happy to use it, especially when she toured Europe and met crowned heads – many to fall in a few years – among them Tsar Nicholas, who was most impressed by her flying and gliding and called on his officials to increase support for Russian pilots.

Back in France she took part in a pylon race, competing against Louis Blériot, among others. It was a hard-fought contest and, at one point, as another pilot overtook her, she found herself caught in the wake of his propeller. The *Daily Mail* correspondent who was present at the race described what followed: 'Amid the cheers of the crowd I noticed her machine was pitching rather dangerously. The airwoman, who was wearing her becoming flying costume of a white woollen motor-hood, white sweater, grey divided skirt with black stripes, cut very short, and brown stockings and shoes, rose to over 200 feet as her machine rounded the pylon . . . but her turn was wide and she seemed rather uncertain of herself. The cause of her fall is still the subject of animated discussion but, whatever the reason, her aeroplane suddenly dipped earthwards and, striking the ground with terrific force, collapsed like a cardboard box. As she realised she was falling, Mme de la Roche uttered a cry which was distinctly heard. A soldier was the first to arrive at the wreck . . . Mme de la Roche was lying still beside her machine, her face covered with blood.' Later, in hospital, she declared she had been brought down by the backdraft of another machine. 'The man who did that to me,' she said, 'was a bandit. He should be banned.'[1] The offending pilot

---

[1] *Daily Mail*, 9 July 1910.

was hauled up before the race committee and cleared of all blame; the dangers of being caught in another craft's slipstream were not really recognised but disturbed air-flow was going to be, and still is, a perennial problem in aviation as the skies become more crowded.

Raymonde's injuries were terrible but her will to get out of bed and back into a plane was more than equal to them, and after a convalescence of two years, supported by Charles Voisin, she was flying again. It was a tragedy of a different sort which occurred shortly after this and led the press to hint that *La femme-oiseau* was being pursued by the malevolent hawk of fate: out motoring with Voisin, their car was involved in an accident and the great French designer and aircraft builder was killed outright. Raymonde was injured, though nowhere nearly so badly as before, and was back in the air within months. She was not, however, flying a Voisin plane, since unlike his dead brother, Gabriel Voisin did not and had never liked the ambitious actress – or perhaps he wished it was his bed she'd been sleeping in; either way he was not inclined to further her career. She found new sponsors and a new plane, and in 1913 she won the Coup Femina, for the woman pilot who had achieved the greatest air distance during the year. Accepting the award, she said it was to the air she had dedicated herself and that she flew 'without the slightest fear'.

During the Great War she was unable to find work flying in either a civilian or military capacity, and perhaps, given the attitude of France towards its women (sacred duty to produce more boys for the cause – in fact, *to be* the cause) that was not so surprising. After the Armistice she returned to flying, capturing the women's altitude record for France in 1919. It was taken back to America by Ruth Elder and Raymonde began looking round for a more powerful craft, something that would utilise the lessons of wartime flying. In the course of her search she went up in an experimental plane with a test pilot. Something went wrong and the ship crashed on landing. This time Raymonde de Laroche didn't survive her injuries.

There was no lack of successors to the dead airwoman in France. The nation had taken to the air with an enthusiasm that far exceeded even that of the Americans in those early years. Perhaps it grew from the tradition of motoring, or was a natural successor to the ballooning craze which had held the nation in its grip for over a century; or it may have been due to the French obsession with all things American. There was also a difference in attitude: French male pilots did not seem to feel as threatened by women in the air. Eileen Lebow suggests, in her seminal study of early women aviators, *Before Amelia,* that the very rigorousness of the training for both sexes resulted in a mutual sense of respect: if a woman earned her licence then she had every right to fly. There was also the old male fascination, national cliché though it undoubtedly was, of a pretty woman in a balloon or a plane; and, in the flying classes at least, there was a different social–sexual attitude towards women. One can't imagine many Frenchmen thinking society would flare up in a Bolshevist Armageddon as the result of allowing a woman to climb into a plane, whereas in England the vocal majority considered the whole idea about as attractive as being trapped in a small lift with Oscar Wilde. One influential industry commentator stated the facts as he saw them: 'There is no place for women in the air.'

Cycling and motoring were not considered to be the sort of thing an English girl should be doing, especially after she was married and settled. And when her husband encouraged her, it was almost too much. But the English establishment has always been able to absorb what it couldn't stop, just as long as those involved were the right sort. And Hilda Hewlett undoubtedly was. Eccentric, yes, but the daughter of a Church of England vicar; an outdoor girl in a Joan Hunter Dunne mould, but also a young woman who thought about the world around her and trained as a nurse and went to art college. She travelled around the Continent and spoke German and French well; she had little time for 'society' as such, and saw her

Christian duty as being a matter of practical intervention; there was never anything mystical about her character. When she was 24 she married Maurice, who was something – but not too much – of a bohemian himself and, commendably, didn't see the need for the word 'obey' in the marriage service. He wrote plays and historical novels, among them *Richard Yea and Nay*, about Richard The Lionheart, and *The Song of the Plough*, about an Englishman's love of the land. He was also to produce some moving poetry during the Great War.

Hilda and Maurice settled down to a vaguely unconventional (in a new Arts Club sort of way) married life. He became a top-ranking civil servant and began to gain a reputation as a writer; she had two children, designed and redesigned the interior décor of their house, became a noted cyclist and got caught up in the motoring craze. She taught herself to be a more than competent mechanic and then, one day in 1909, Hilda Hewlett met a Frenchman called Gustave Blondeau, who loved flying; and she, who loved enthusiasm, decided to take a look at this new thing and see if it was just a craze or something longer-lasting. Accepting Blondeau's invitation to a flying show in the north of England – the two drove an open topped auto from London to Blackpool, no mean feat in those days – she was deeply impressed by what she saw and felt she either 'wanted to cry or shout'. She agreed absolutely with her new friend that the future lay in the sky.

They decided to form a company, start a school, in fact do anything they could to get hold of a plane and learn how to handle it. Hilda was able to raise enough cash to buy an aircraft and the pair settled on a French model – there wasn't a British craft on the market – made by Charles Voisin's old partner, Farman. Blondeau learned to fly the plane in France, it was then taken apart, crated and shipped to Brooklands motor racing circuit in England where, Hilda had discovered, there was flat ground in the centre of the track which could be hired and turned into a flying field. Here, they rebuilt the plane and she set about learning to fly it. Local motor racing enthusiasts were less than encouraging, disparaging the plane, its engine, the pilots and their idea of setting up a flying school to pass on their skills. Their prognostications proved false. Hilda achieved her licence in August 1911; she was 47 years old and the first Englishwoman to do so.

The flying school, with its boast that no pupil had ever crashed or even damaged a plane, flourished; and the boast was no idle one – they had one plane and used to watch every lesson with trepidation in case it was damaged. They decided that, to spare their nerves, if not their purses, they would invest in a second craft and the equipment to keep it in the air; they were the first people in Britain to import and use oxyacetylene welding equipment which, as Hilda said, soon paid for itself since other flyers and motorists, too, were soon coming by to get their welding work done. The school's pupils became part of a fraternity of flyers, often camping beside the airfield during their holidays and helping out with the mechanical work, the cooking and washing up.

Hilda was invited to speak to various august bodies in the British flying world and told them all that the country was lagging far behind in the development of a home-grown aircraft industry. One of those who heard her and took note was the young Tommy Sopwith, who went on to design a whole range of famous planes bearing his name. Never a woman to talk when action was possible, Hilda took her own advice and, when the First World War started, went into partnership with Blondeau to open an aircraft factory located first in London and later in Bedfordshire.

The company survived a brush with early union action – Mrs Hewlett was not sympathetic to the workers' requests – and ran successfully until 1919, when it was closed down. Her son, whom she had taught to fly, became an aviator; her husband Maurice, buried in his writing (he had a play opening on Broadway in 1918) and his civil service work, took little interest in Hilda's new career and indeed, his unconventionality stopped short of allowing him to visualise women in the air. Hilda saw no reason at all why they should not fly. Husband and wife drifted apart. He died in 1923, and she moved to New Zealand.

Lilian Bland, a young Irishwoman who covered sports events for the newspapers and photographed birds for her own satisfaction, became fascinated by the idea of powered flight after a relation sent her a postcard from France with a picture of Louis Blériot's Channel-crossing plane on it. Always a practical person, she decided to use the workshop her uncle, General Smyth, had installed at his house outside Belfast to build her own plane. This wasn't so unusual at the time, particularly in Great Britain, where, as Hilda Hewlett was forever pointing out, aircraft manufacturing had not yet established itself. Looking at the Wright brothers designs, Lilian constructed a series of gliders of increasing size until she had a craft she thought would be big enough to support the weight of an engine. She tested its lifting power with four local policemen who were persuaded to hang on to the wings while the gardener's boy worked the controls. The glider flew easily – more easily once the constables let go and the craft sailed away, with the anxious boy trying to bring it safely to the ground.

Lilian was satisfied and ordered an engine from England, which was fitted the moment it arrived. There was no petrol tank, so the aeronaut jury-rigged one, using an empty whisky bottle. There followed a series of tests which were reported in *Flight* magazine with mounting excitement as the craft taxied then began to make short hops. Lilian couldn't believe it was happening at first; she had to run back and check the point where the wheel stopped leaving tracks in the grass to prove to herself that it was true.

'I have flown,' she wrote to *Flight*, 'All this time I have been learning my engine and getting things shipshape. Then for five weeks we had fearful weather and are having it still . . . at last, on Wednesday, I got her out again and she rose to 30 feet, which I carefully measured. It was dead calm so there was no wind to help her . . . The aero engine is splendid, although I frequently have fights with it to get it to start . . . I am naturally awfully pleased, having made and designed her myself. It is a very small but promising start anyway.'[2]

As she became more experienced *The Mayfly*, as the craft was christened with a touch of whimsical Irish humour, was remodelled again and again until she felt the design was ready to be offered to the public. There was no response; her gliders continued to sell but there was just no demand for the do-it-yourself ship from Belfast. Glad to have done it and flown, Lilian was not particularly upset; she felt that the plane was not really up to commercial standards. She offered it for sale and the Avro Roe engine, according to historian Peter Lewis, ended up in London's Science Museum. Lilian herself turned to motoring and later, like Hilda Hewlett, headed out for the colonies, in her case emigrating to Canada.

Germany, preoccupied with dreams of Empire as the twentieth century began, was not slow in coming to see the potential of air travel and air power. In 1903, Karl Juthe was making short hops in his experimental monoplane but was unable to cover more than 60 metres. The first German to fly successfully with a German engine was Hans Grade, on 30 October 1909. He covered a figure-of-eight course around two pylons. In the same year, Count Ferdinand von Zeppelin inaugurated the world's first passenger service, operating airships rather than planes, and it was in this direction that much German aeronautical thinking turned. The service conveyed 34 000 passengers during its four-year life, before the Great War put a stop to commercial airship transport and opened the throttle on the development of aeroplanes.

Amelie 'Melli' Beese was, like many women pilots, not afraid to be unconventional in pursuit of her own ambitions. The child of a respectable middle-class family, she was born in 1886 and grew up in the overpowering atmosphere of

[2] Letter from Lilian Bland, *Flight*, 10 September 1910.

biedermeier culture; she, however, was not made of the stuff that made mothers for a greater Germany. Melli preferred the arts, particularly sculpture, which was about as far as the family was prepared to let her go – though she did have to travel to Sweden to study since no German school would take an unmarried woman. When she graduated, she returned home and opened her own studio. In the meantime, the news of the Wright brothers' flight had set something working in her mind and she decided she too would fly. Not so easy, of course, in a country still discovering its national identity, where the class structure was even more rigid, because it was established on far shallower foundations than in Imperial Great Britain. However, she did at last gain admittance to a school linked to a company called Ad Astra, established by an international grouping of flight enthusiasts. The set-up was rather like the Hewlett–Blondeau school in England and she became part of a student society that sprang up around the workshops, the field and the local coffee houses.

Melli gained her licence in 1911, but only after suffering an accident that resulted in hospital treatment, a forced convalescence and then a move to another flying school. She also suffered terrible guilt over her father's early death, believing that her wild lifestyle contributed to his heart attack. That is unlikely, but it does give an early indication of her tendency to look on the dark side of life. Whatever the truth of it, her father's will helped Melli to fund her career and she began racing and trying for records, setting a German women's endurance record of 2 hours in the autumn of 1911. Her passenger on this flight, a young Swiss professional, paid tribute to her skill, stating that there she wasn't just a pretty face (and she was a very beautiful woman) but really could fly. Others, apparently, were less supportive and there are stories, unproven but convincing, that her early flights were the subject of sabotage attempts by men who felt her presence was an insult to . . . almost anything Germanic you care to name. Certainly, she was always careful to check everything from spark plugs to wing wires before she took off on a flight. It's impossible to know the truth of the matter – she did suffer a lot of crashes, but then so did just about every other early aviator, man or woman. Flying was dangerous; planes and their engines were undergoing constant development, and not every step was a step forward. It was only the comparatively low speeds, particularly during take-off and landing, that ensured the survival of so many pilots. Dutch pilot and designer Anthony Fokker lamented that just about every airfield he knew had been irrigated with the blood of his brother pilots. His memory, he said, was one long obituary list.

In 1912 the 'Flying Girl' (the German press couldn't quite match the élan of their French cousins) set a new height record by accident and became the press 'sensation of the week', in her fur coat and slacks. She was shrewd enough to realise that celebrity would mean nothing if she didn't capitalise on it, and she began planning a flying school of her own. This would enable her to finance her own activities as well as popularising the sport in Germany. Her celebrity had a more negative side to it as well: crowds would spill over on to the runway whenever she appeared, making it dangerous to attempt landings, but this small danger paled beside the growing jealousy of her male colleagues: they didn't like this flying girl coming between them and their thunder. Jokes were played – deeply serious German jokes intended to end in at least a broken leg, if not the twilight of her career. She redoubled her checks, put up with the hostility and took the precaution of flying alongside two male pilots she could trust: the young industrialist Alfred Peitschker and the Frenchman Charles Boutard. Both men immediately, predictably, fell in love with her. Peitschker proposed marriage; Melli proposed friendship. An air crash solved the dilemma for the young man; he was killed testing an experimental plane.

Knowing the guilt she'd felt over her father's death, it is likely that Melli was plunged once more into depression by this new attack on her conscience. Did she marry Boutard on the rebound? Who knows – he was happy with the arrangement, she was too, it seems, for the couple set up a flying school and a design business. Work was hard and non-stop, but slowly the school began to pay its way, though it was never going to make either of them rich. Their first plane came off the drawing board and into the factory, to be shown in 1913. It seemed their future was assured as a second model of the Beese plane was unveiled and tested with encouraging results. Melli began to think abut designing a flying boat, but everything was put on hold by a pistol shot in Sarajevo. As Germany mobilised for war, Boutard and Melli were arrested as enemy aliens – obviously absurd in Beese's case, but those of her countrymen who had resisted eating their revenge hot, now got a chance to enjoy the dish cold. Boutard was interned, Melli had her licence rescinded and her factory, her plans, everything she owned taken from her, including, so Melli biographer Mike Graham asserts on his website, the design of the Beese–Boutard monoplane which reappeared during the great war as a Fokker fighter.

After the defeat, as Germany floundered among the revolutionaries and the Freicorps, Melli found the released and seriously ill Boutard and the two of them tried to pull their lives and their business back together again. The lives somehow couldn't be fixed; the couple separated. Boutard went back to France where he was arrested as a collaborator. The business was impossible to resuscitate, but with lawyers on the case, Melli was able to gain compensation, which came through in 1923, just as Weimar sank into the pit of inflation. The lawyers took a large bite out of what little was left over.

Melli was determined to get back into the air. Around the world aviation records were being shattered every month. A new, commercial industry was coming into being: in Germany, a daily passenger service now operated between Weimar and Berlin; London–Paris, Paris–Strasbourg flights were in operation; internal services were springing up all over America. Melli was certain that if she could get her licence back and a few more miles in her logbook, she could restart her career. In 1925 she began flying again but experienced a difficult landing in which the plane was a write-off, though she walked away without any external injuries. Inside was a another matter. Like the later British flyer Amy Johnson, she had always been prone to fluctuations of mood and though she possessed a natural strength of character that had pulled her out of depressions before, she had been through a ruinous war and was now coming to uneasy terms with a bankrupted peace. She couldn't find the will to go on, and shot herself on 22 December 1925.

Harriet Quimby was the first American woman to receive her pilot's licence though not, according to Matilde Moisant (officially the second) the first to actually qualify for it. Moisant told Bobbi Trout that *she* was the first to complete the training and go solo but because Quimby was going to have to earn a living in the flying world and Moisant was not (she was in the aircraft business with her brother) she let her friend have the advantage of being first. In fact, neither was the first American woman to fly; Blanche Stewart-Scott did so at the Glenn Curtiss flying school at the beginning of September 1910. This was by no means an easy task. Curtiss, like a lot of men, was against women pilots. He thought they'd never master the intricacies of doing one thing with their hands and another with their feet (obviously forgetting a million piano players); he was also concerned that their inevitable crashes would raise public concern and put the cause of flight back by many years.

Stewart-Scott had recently driven a car from New York to San Francisco to demonstrate, for her sponsoring motor company, that driving was so easy that even a woman could do it. Persuading Curtiss wasn't easy, but she managed to do that too and he put her through an intensive course on aeronautics – three days – and then allowed her to practise taxiing in one of his planes. He took the precaution of fixing the throttle so that the engine wouldn't deliver enough power to allow Scott to take off; however, the elements were on her side and one afternoon a gust of wind gave her just enough lift to get off the ground and prove to Curtiss that not only could she take off, she could also land. She could also pull the wool over his eyes: she had removed the lock on the throttle to gain the necessary engine power. She soon became a member of his flying troupe and performed with the Curtiss Circus all over the country. Curtiss, perhaps awed by her duplicity, decided never to teach another woman to fly – and he kept his word.

At the end of 1910, Bessica Raiche flew in a machine built by herself and her husband Frank. Neither woman gained a licence, which was not actually necessary at this time: in 1916, Blanche Stewart-Scott left the air-show circuit stating that there was no way a woman could earn a proper living in flying, and went on to film production, acting and writing; Bessica Raiche, having proved to her own satisfaction that women could fly, concentrated on aircraft design and production and then went into medicine.

Harriet Quimby had started out as a reporter, and a very successful one too. After a childhood variously described as wealthy and poverty-stricken and which was probably an amalgam of both, she began writing short pieces and drama criticism for Los Angeles papers before going to New York and bluffing herself into a job on *Leslie's Illustrated Weekly*, one of the top news magazines of the time. She wrote hard-hitting articles on female suffrage, delivered cutting judgements on the theatre, and more than held her own in the male-dominated world of New York newspapers. She was covering the Belmont Park air show in 1910 at which a daring young aviator, John Moisant, won what may well have been the first pylon race around the Statue of Liberty. The flyer was killed a few months later, before Harriet could get to know him as well as she would have liked, but she did become friends with his sister Matilde and, through her, met a second Moisant brother, Alfred, who ran the Moisant flying school. Never a woman to back down from the chance of a new experience, Harriet signed up for a course and Matilde joined her.

After receiving her licence, Harriet was invited to join the Moisant International Aviators, an exhibition flying team which appeared all over the country. She didn't really need the publicity; as a newspaperwoman she knew exactly how to get the most out of her flying activities. She commented that after writing about others for years, it was a pleasure to sit back and read about herself. This was slightly disingenuous since she was managing her own publicity with a professionalism G. P. Putnam might have envied. A large part of the Dresden China Aviatrix's public persona came from her theatrical flying costume of plum-coloured silk blouse, bloomers and turban, laced boots, leather coat and goggles; an equally large part of her professional self depended on a natural ability in the air combined with a coolly calculating mind and real sense of pushing the envelope, as the Edwards Air Force base boys would later call it.

Having carved out a place for herself in the tough world of journalism, she was less than pleased to find that the various bodies controlling flying in America were doing nothing to advance the position of women pilots. In many cases they were not even allowed to compete, plane against plane, with the men. Apart from anything else, this meant the opportunities to race and the prize money available were far less than male pilots could win. It was clear the authorities were using the economic weapon to maintain the status quo. Harriet saw no reason at all (apart from male prejudice) why women should not earn a good living from flying. Passenger carrying, mail delivery and air photography were all areas she considered as more than suitable and she was determined to push not only her case but a case for all women through her journalism. She also realised, as a good hack, that publicity wouldn't do her any harm either and began to plan a flight across the English Channel. Louis Blériot had been the first; the late John Moisant had crossed from Paris to London in 1910; she decided to fly from England to France in April 1912.

In England she got in touch with the *Daily Mirror* newspaper and sold her story up front. She was concerned in case word leaked out in advance and some other woman flyer jumped off in her place; she conducted the negotiations and all her other plans in as much secrecy as she could manage. She crossed to France, met Blériot and hired a plane which was crated up and sent across to Dover, where the press were gathered, without knowing why, to await the arrival of the Dresden China Channel Crosser. Among those waiting was Gustav Hamel, a British flyer she may have met through her interest in mail flights. He had inaugurated the British airmail system a few years earlier, flying letters across London and ferrying thousands of congratulatory telegrams to Windsor on the occasion of King George V's coronation. He had recently flown the Channel and, as a proponent of women flyers, was happy to give any support he could to Harriet. He was also keen, it appears, to get into her plum-coloured silk bloomers, since, on the grounds that the flight was extremely dangerous, he offered to don said garments, make the flight disguised as Harriet and switch back in France so she could claim the credit. An odd and apparently genuine offer which Harriet, quite naturally, rejected. She did, however, take Hamel's advice on using a compass (which she had never done) and listen to his warning that if she were blown more than five miles off course, she could end up over the North Sea where her Blériot monoplane would be subject to vile conditions and, in all probability, come down. He gave her a hot water bottle which he 'insisted on tying to [her] waist like a large locket'.

She turned on the ignition, the prop was spun and the engine roared. As she built up the power, Hamel and five others grabbed hold of the plane, which had no brakes, straining until Harriet had reached the required revs. She waved and they let go; the craft sprang forward and took off at 5.30 a.m. She gained height and then rendezvoused with the film cameras over Dover Castle. The day was misty and she could no more see them than they could photograph her. She realised she would have to make the flight by compass, while timing her progress, hoping the weather would clear over the Channel. It didn't but, as she neared what she hoped was the coast, rifts at last appeared in the mist and the morning sun began to shine through, down on to 'the white and sandy shores of France'. She began to descend, looking for landmarks. She saw none that she recognised as being near Calais, but it was France, she'd made it; the hot water bottle was as cold as ice and she was freezing so she landed, to be met by a crowd of fishermen. She *had* done it – but as she knew, she hadn't really done anything until the press arrived, cold-foot from Calais and off the *Daily Mirror* tug which had been shadowing her, to record the event in words and pictures.

It was stunningly bad luck that as she was going up, the *Titanic* was going down, in far colder seas than she'd flown over, and the newspapers of the next few days were full of tragedy rather than triumph. But she was not put off; she published articles on the flight in America and Europe and was, by those who realised the scale of the achievement, granted real respect. She didn't have long to enjoy it.

In July 1912 she was back in the States at the Boston air meeting demonstrating a new Blériot monoplane. Her manager William Willard asked if he could come along for the flight. Harriet was quite happy to take him, even though the Blériot was not a well-balanced plane (a sack of sand was required to keep it level in flight) and Willard, who was no lightweight himself, might cause serious stability problems. She was confident in her ability as a pilot and told a concerned friend that she had no intention of ditching in the bay, since she was like a cat and hated water.

The Blériot took off after Harriet had warned Willard not to move around at all; though, oddly, given the care she usually took to check 'every wire and screw' on the ship, she had omitted to strap either herself or her passenger into their seats. Nobody knows quite how it occurred but the whole crowd saw what happened next: at about 2000 feet, as Harriet was turning the plane before coming in to land, Willard, for some reason, shifted his position; the Blériot bucked and he was thrown out. For a few moments, Harriet appeared to have regained stability but the craft flipped again, this time throwing her out. She fell into the shallow waters of the bay and was killed instantly. The New York papers opined that her death should make it clear that women had no business in the business of flying and should get out of it.

Right about this time a very determined young woman was ignoring the advice and doing the exact opposite and getting right into it. Katherine Stinson had trained as a pianist but decided to become a pilot. She was one of four children, all of whom became involved in aviation. Her parents split up, apparently quite amicably, and her mother, Emma, took the family off to Mississippi and then Arkansas, where she established herself as the publisher of a series of highly successful city directories. Money was never really a problem for the Stinsons; neither was a lack of purpose or convention. Emma Stinson saw no reason why her daughters shouldn't have the same upbringing as her sons and when Katherine gave up the piano and picked up the idea of flying, she was happy enough to let her learn, once they could find a pilot who would teach a girl.

They did, but only by waving a pile of dollars in his face and, during her first lesson, after taking the Wright Flyer up solo, Katherine realised that while the instructor had her money and she had the aeroplane, she didn't have the slightest idea of how to land it. Bellowed instructions from the instructor, who could see his living ending up in a pile of broken struts and flapping canvas, got her down. She loved flying, and spent every moment she could at the airfield, earning her licence in July 1912, just 16 days after Harriet Quimby's fatal crash.

Even with a licence, Katherine found herself, as so many women had, with problems: male pilots on the field would pass comments about so slim and pretty a girl being crazy enough to want to learn to fly (there was always something about the combination pretty-and-flying that brought out the worst in the boys) and got extra laughs out of watching the five-foot, 101 lb aviatrix obsessively cleaning down her plane. They called her an old maid and told her that oil and grease wouldn't harm the wires and joints one little bit. However, when the thick, black grease was eventually cleaned from the various parts, Katherine discovered that her second-hand plane was a death-trap. Frayed wire, cracked couplings, missing bolts: any one of them could have ended a flight in a catastrophic manner. Her fellow pilots began to reassess the newcomer and think about the maintenance of their own craft. It wasn't long before most of them were scrubbing and scraping away at wires and wheels and engine parts. As Katherine said, if your car goes wrong on the road, you can pull over, get out and look under the bonnet. If your aeroplane breaks down in flight, no way can you get out and sit on a convenient cloud while you check the motor.

The Stinsons being the Stinsons, the first thing they did was form a company to develop their aviation futures. They had a plane, the Wright Flyer B that Katherine had cleaned down, and now they thoroughly reconditioned it themselves, learning as they went. Katherine got herself a manager who reconditioned *her*, taking a few years off her already tender age, utilising her slim, slight figure, cute looks and tumbling brown locks to relaunch her as the Flying Schoolgirl!

She was a sensation. There was no need for elaborate stunt flying; performing at country fairs, spectators were thrilled just to sit next to her in the plane without ever leaving the ground. And when she actually took off and flew round the field and low over their heads, performing the very simple stunts that were all her elderly plane could manage, they were ecstatic.

But the public is fickle and tastes rarely stay simple – her audiences began to demand more, and she was happy to oblige. The most famous of the stunts she introduced was looping the loop, though she had to wait until 1915, when she was flying a more stable plane with a better engine, before she could attempt the manoeuvre since the stunt often resulted in engine stalls at the top of the loop; and no pilot wanted to find themselves in an old ship if that happened. One possible solution was to loop as high as possible, thus giving the aviatrix time to restart her motor before the plane spun into the ground. As it turned out, she was to need every foot of the extra height when she tried serial looping for the first time at a Sunday air show. She had accomplished three loops perfectly but at the top of the fourth the engine stopped. It wasn't a stall; one of the valves had broken, and there was no chance at all of restarting the engine. Katherine recalled later that in the sudden silence after the engine stopped she experienced a feeling of utter horror but realising that once the plane lost its balance and began to cartwheel down, she would be irretrievably lost, she pushed the stick forward, bringing the nose down, so the ship began a straight dive towards the earth. It approached at terrifying speed, but she knew that if she could pull the stick back there was a chance – just a chance – she would be able to lessen the angle of descent enough to land. She did, and sat frozen in the cockpit as the ground crew rushed forward and the whole crowd let out its breath in one enormous, relieved (disappointed?) sigh. Her response, once she could trust herself to stand up and speak without bursting into tears, was to make plans to include the 'death dive' in her show in future. She was persuaded this was a touch of unnecessary bravado, though she did add a roll at the top of her loops in later exhibitions. She was the fourth pilot and first woman to loop the loop, and experienced a certain amount of jealousy from male pilots who had yet to achieve the feat.

In 1916 she set off on a tour of the Far East, doing flying exhibitions in Japan, where she attracted huge crowds and introduced the idea of flying to many Japanese women, who were encouraged to sign up at flying schools. Once Katherine had left, the women were encouraged to return home by their husbands and teachers. She went on to Shanghai and Beijing where the crowds were vast and positively dangerous to woman and machine as they overflowed the public enclosure in their enthusiasm to actually touch the wonderful creature.

Back home in America, she began looking around for some way to earn a living which wasn't as dangerous as stunt flying; she decided on the nascent airmail service and applied to join. She had some experience of flying mail in Canada and thought the new US service should be taking account of women flyers by now, and would be happy to use her talents. She was, after all, the best woman flyer around. She was also wrong – predictably, her application was shunted around offices, sent here there and everywhere but not by airmail, and finally put on a spike somewhere, where it would be dealt with in the future. This wasn't good enough for Katherine, who went straight to the top and spoke to the Postmaster General. He supported her application and enrolled her as an airmail pilot. There were further attempts at discouragement and even sabotage by members of the service, but nothing could stand before the determination of the 101 pound steamroller. She once told her mother that if you let other people decide what you should do, then you'd never do anything.

When America entered the First World War, Katherine decided to leave the airmail service and do something more practical for her country. She was not allowed to fly as a fighter pilot, so she toured, collecting pledges for the Red Cross, raising almost $2 million, and setting up a distance record as a by-product of the charity drive. She also helped her younger sister Marjory who was training Canadian fighter pilots in Texas and was known, in consequence, as the Flying Schoolmarm (you have to hand it to the hacks, they really are irrepressible!). Then, wanting a more active role, Katherine decided to go to France and drive an ambulance. Here, in typical fashion, she pushed herself up to and beyond the limit and, in an exhausted state, contracted tuberculosis. She was invalided home, where she was to spend six years fighting to recover her health. She won, but she never went back to her career as a pilot.

The role of women pilots in war presented big problems to the various military authorities involved. There were exceptions

– the Russians, Tsarist and Communist, were always ready to set an example of equal treatment, whether in the air or in the *gulag*; Turkey, in the case of Sabiha Gokcen, an orphan adopted by Kamal Attatruk, produced the world's first woman combat pilot but on the whole there was always something about the idea that made the male military mind uncomfortable. After all, it was home and hearth that they were all defending, and to have homemakers and hearthkeepers right there beside you, made the whole thing look faintly ridiculous.

French pilot Marie Marvingt proposed an air-ambulance service as early as 1910 but her proposals fell on the ears of the French military, which were as near deaf as makes no difference. Once the war started, a few planes were used to evacuate wounded men, but there were so many wounded and so few planes that the practice soon fell out of use. Helen Dutrieu, another Frenchwoman, is said to have volunteered as a military pilot when hostilities began. She was posted to a reconnaissance squadron, though later is said to have flown combat missions against zeppelin bombing raids over Paris.

The Russian Princess Eugenie Shakhovskaya was rumoured to have been a combat pilot in the early years of the war, though this is unlikely. She had learnt to fly in Germany from the Wright Company's pilot and soon gained her licence, the second Russian woman to do so. Little is known of her career beyond the fact that she flew with male pilot V. Abramovich, possibly also her lover, who was killed when testing a Wright Flyer with Eugenie. She was injured but survived to fly again domestically and later, during the war, either as an artillery spotter or reconnaissance pilot. At this point the story begins to sound like something out of Catherine the Great's time, with the lusty Eugenie attracting dozens, even hundreds, of moustache-twirling Tsarist officers to her bed while squandering her fortune on champagne and wild parties, no doubt with gypsy singers and *troikas* harnessed outside for a quick dash to the country *dacha*. Later, it was said, she became a spy for the Kaiser, selling the secrets her sated lovers whispered to her across the silken pillows. Betrayed, or discovered, she was arrested and imprisoned. A military court sentenced her to be shot by firing squad but Tsar Nicholas stepped in and, either because of some family connection or perhaps another sort of liaison, pardoned her and had her sent off to life imprisonment in a nunnery. Here the tale threatens to turn into a novel by Bulgakov, as Eugenie was freed by the Red Army, discovered the error of her aristocratic ways, read Marx from cover to cover (surely the only utterly unlikely part of the story) and joined the Secret Police in Kiev. Here she surrendered to the charms of Lady Morphine, became an addict and was promoted to the position of official Bolshevik executioner for the city. Finally, in a drugged haze, she shot one of her assistants with her Mauser automatic, at which point her fellow reds turned on the deadly princess and shot her down. Somehow one hopes it is all true.

The ground, or rather, the sky had been marked out – the vapour trails drawn ready for the 1920s and 1930s, the great age of flying, when pilots would write their names across the blue and into the record books.

4
Death of an unknown woman

*I*n Britain, the summer of 1939 was one of the best in years, clear and unclouded except by the threat of a European war which an ineffective government was resolutely refusing to see as any kind of threat at all. Life went on and along the City Road in London's East End, the traffic was as heavy as on any other day and the crowds as busy. The passengers on the rattling tram heading towards Stepney were looking the other way as resolutely as the government, trying in that very English manner not to notice the woman who seemed to be drunk. She looked respectable, but might have been Irish and though she was causing no trouble to speak of, the very juxtaposition of 'woman' and 'drunk' made everyone uncomfortable. It wasn't done; it wasn't right; and her fellow passengers were relieved when the tram pulled in to a stop and she got up and wandered along towards the exit platform. At that moment the tram bumped as it hit the points and she lost her footing and fell, hitting her head on the open platform. She didn't try to get up, but lay where she fell. The guard hurried to a nearby phone and summoned help. An ambulance soon arrived and, still unconscious, the woman was taken to St Leonard's Hospital in Shoreditch where she was pronounced dead on arrival. An examination found the cause of death to be a fractured skull sustained after the fall on the bus – it was noted that the skull bore signs of a second fracture sustained perhaps ten or so years before. Beyond this and the fact that her clothes, though old, were of good quality, that she was carrying only a few pence and that she was aged around 40, nothing was known. Who she was, where she was going by tram that day, where she'd come from, were all a mystery. It was just the death of an unknown woman one day in the summer before the war.

She was born 43 years earlier in Limerick. Her father had died three months before the birth, her mother very shortly after it. She was looked after by aunts and grandparents and her name was Sophie Catharine Theresa Mary Pierce-Evans. The family was prodigal in names but poor in cash – at least, according to Mary, as she always called herself, though she was telling the story years later, when circumstances had enhanced her natural conviction that a pinch or two of fiction always added something to the stew.

Mary was educated in and around Cork, changing schools often since her aunts disapproved of sports for girls and moved her on as soon as she began to excel, as she always did, in the school hockey or lacrosse teams. She also excelled in bending and sometimes breaking the rules. In an article she wrote years later for *School-Days Magazine*, she recalled the many scrapes she got into, flouting the rules, accepting every dare that was offered and, most of all, making up elaborate and thrilling tales that peopled the ancient buildings in which she and her schoolmates studied with bloody villains and howling prisoners covered in blood. On one occasion she was able to fool her headmistress with a story that got a friend released for the afternoon to visit an imaginary garden party.

She went on to study at the Royal College of Science in Dublin, where she took her BSc in land management. Or not – there is no record of her academic achievements, and this might be due either to a lack of achievements or a lack of records, as the archives were burnt when Dublin Castle was destroyed during the Easter Rising of 1916. What we do know is that Mary Pierce-Evans had started out studying medicine but had been unable to complete her course and had switched to land management with the intention of farming in South Africa.

Her interest in athletics, which had so bothered her aunts, had grown; she was beginning to become an advocate for the right of women to run or jump in any competition they wanted to enter – something that was not acceptable to the Olympics committee who rejected an entrant to the 1912 modern pentathlon on the grounds that she was a 15-year-old girl.

During the Great War, Mary joined up and trained as an Air Force dispatch rider, enjoying the chance to roar around the country on a powerful motorbike. She continued to participate in athletics both as competitor and, increasingly, an administrator. It was becoming clear to her that two things mattered above all if she, and women generally, were to make the progress they wanted after the war ended and the men came home: money and public attention.

In America, would-be pilot Bobbi Trout pumped gas near the Anaheim Ditch to earn the money for flying lessons. In England, where she now lived and worked as a stenographer, Mary Pierce-Evans married the 76-year-old Major Elliot-Lynn. It's hard to see the marriage as anything but a matter of convenience for the young, fit and intensely ambitious Mary. Elliot-Lynn's position and money gave her a plinth on which to stand, from which she could attract public notice: 'Ah, yes, Mrs Elliot-Lynn, decent woman, good marriage, talks common sense.' Cynical, maybe, but also appealing to something in Mary Pierce-Evans' character that loved to tweak noses, to surprise her audience with a sudden twist in the story or even to shock them with a trick ending.

The strategy, if strategy it was, worked and she was able to make her voice on athletics heard loud and clear. She was picked as a member of the Great Britain Olympic Squad in 1924, and had a track (actually field, she gained a medal in the high jump) record to back up her words. She was also a founder member of the English Women's Athletic Association, which came into being in 1922. In 1925, as Mrs Elliot-Lynn, she presented a series of papers on women's sports to the International Olympic Committee, which didn't see fit to elect a woman member until 1981, and went on to publish these as a book, *Athletics for Women and Girls: How to Be an Athlete and Why.*

She was fast becoming a highly influential member of the athletics world when problems began to arise in her marriage to the Major. He proved to be rather more dashing than anyone suspected when he deserted his young wife and upped and dashed off to Kenya. She later said, 'We were very happy. I did everything I could to keep my home but he was a wanderer and went abroad.' Sympathy was general, at least general enough for her to keep her position within the athletics establishment, but money was short and she was not best pleased at being deserted by a 77-year-old. She set off after him, travelling steerage to Kenya, no very pleasant experience in those days of multiclass ships. Once there, she set off to find the wanderer, who had taken considerable trouble to cover his tracks.

After months spent searching through the white colony in Kenya (since it was small and very exclusive, not to mention amoral, one suspects that friends were hiding the ancient runaway), Mary eventually found him and asked for a divorce. He refused. She pleaded. He refused. She begged. He refused. There must have been something annoying him an awful lot, hurt vanity, perhaps, but nothing Mary could do would bring him round. She went back to London and took up shorthand typing. She also took up flying which, she said, was really absurdly easy to learn. She thought, after trying a first flight, that 20 lessons of half an hour each were enough to teach the average person all they needed to know.

With Elliot-Lynn still an inconvenient husband but at least out of the way at the other side of the world and provided with enough funds to pay for her lessons, Mary threw herself into this new pursuit with all her customary enthusiasm. She gained her A licence as a private pilot in 1924 and started a second course to earn her B licence, which would allow her to maintain engines and fly commercially and for profit. It was only then that she discovered that the International Commission for Air Navigation expressly excluded women from flying commercially even if they had passed their B licence. It had only been a couple of years since the Commission had withdrawn its ban against women gaining a flying licence in any way, shape or form: 'The Candidate (for examination) must be of the male sex, must have complete use of all of his four limbs, must not be completely deprived of the use of either eye and must . . . be free from a history of morbid or nervous trouble.'

When the French Under-Secretary for Air noticed this would deprive a celebrated French aviatrix, Mademoiselle Bolland, of her right to fly, it was rescinded, but the commercial ban still stood.

Mary, and a reporter friend, Stella Wolfe-Murray, noted that while the British Minister for Air was making speeches encouraging women to fly, the international flying community was putting barriers in their way. Stella Wolfe-Murray went to America where, as she reports, she met Katherine Stinson, who had taught over 80 men to fly during the war. Mary took up her pen back in England and wrote to the International Commission. She did not mince her words, but attacked the problem of the unspoken but highly influential male attitude towards women's 'main physical disability', directly.

> I clearly realise that the decisions arrived at by your Sub-Commission are due to the very great differences in the nervous and mental systems of some women at certain periods, and it is on examination of a number of women pilots, extended daily over a number of weeks or months that I base my plea.
>
> Having dealt for some years with hundreds of girls and women of all ages and nationalities, I have seen that when a woman is sufficiently fine and healthy to enter the first class of athletics her periods of nervous difference, or her powers of endurance, are imperceptible even to the acute measurements of such instruments as the Reid Indicator, which is accepted by our Air Ministry for measurements of the nervous reactions of intending pilots.
>
> I would like to offer myself as material for any set of experiments that the Medical Commission would care for me to undergo.

It should be said that there is no record of any series of tests carried out by Mary on women pilots for any length of time, but as she was well aware that had they been carried out they would undoubtedly have supported the conclusions she presented. She'd obviously realised that, for many reasons, menstruation exerted an influence over the top brass of the Air Ministry out of all proportion to its effect on women and their lives. The problem wasn't confined to Britain, either; in America, in 1943, Commander Del Scharr was directed by the Army Medical Corps to call her squadron of Air Transport Auxiliary (ATA) pilots together and find out from each when her next period was due so she could be taken off the active flying list; naturally the Army Air Corps received a short, sharp response and dropped the matter forthwith, realising that this was a battle they could not win – whether they ever realised it was a battle they should never have even contemplated is another matter entirely.

Mary Elliot-Lynn was confident of her own ability to sail through any mental or physical examination which the authorities might care to put to her; they didn't seem inclined to rise to the challenge, though, and she began to feel that another broadside might be in order. Then, out of the blue, she was asked to present herself at the Air Ministry in London for a series of tests. She'd been out dancing the night before until 3 a.m. but, as Stella Wolfe-Murray said, in the interests of science and her sisters in aviation, she went at once and passed every test the Ministry set her. The International Commission still would not budge. So she went back and took her B licence anyway, becoming accomplished not only in flying multimotor planes but also, and far more importantly in view of the future, in maintaining their engines.

If flying was hard to get into, mechanics was worse – all that grease, those spanners and nuts, cylinders and plugs! Engines and engineering have always been the gossip of men and they weren't about to let women in without a struggle. When they did, as in the case of Amy Johnson, they turned them into honorary chaps, Amy becoming Johnny. This was not going to work with Mary Elliot-Lynn: she had a right to learn, she would learn and if they didn't like it, then she'd live up to the nickname they gave her: Lady Hell-of-a-din. Of course, she would pay for it in the end – outsiders who go up against the English establishment always pay for it in the end.

Her goal was to present the results of her tests and her experience to the Commission when it sat in Paris in 1926 – the year of the General Strike in Britain. For all her radicalism, Mary was no particular supporter of the British working man and saw the strike as an opportunity to advance her case. She joined a small group of airmen flying newspapers between London and Paris. It was a volunteer mission but was also, quite obviously, commercial flying and the members of the committee could no longer ignore the fact. The ban was rescinded, though the medical committee to oversee the physical testing of women did not have one female member. Back in Britain the Medical Women's Foundation prepared a report on women pilots and submitted it to the Air Ministry which forwarded it to Paris, which resulted in an examination process more likely to encourage women attendees.

In a mood to prove that nothing in the air should be impossible for women, Mary set out to make a parachute jump. The concept of jumping out of planes (balloon jumping was a well established entertainment) was still novel; the Air Force had, only the previous year, allowed aircrew to take parachutes up with them. The argument was that the pilots would take more care of their aircraft if they weren't encouraged to jump out of them at the slightest problem. And, of course, they would all be quite happy to burn or crash with them whenever a real accident occurred. Mary reasoned that safety should always be put first and went up as a passenger to make her jump at the London Aeroplane Club. She climbed out on to the wing and hung onto the struts waiting for the right moment. When jumping from a biplane, it was considered far safer to launch from the wing rather than try to exit directly from the cockpit, where there was a considerable risk of hitting or being caught by the tailplane. However, Mary's plane developed engine trouble and began slipping and sliding across the sky to such an extent that the wing-walker could neither jump nor climb back aboard, and had to hang on until a bumpy landing was made. The next day she went up again and this time, as *Flight* magazine reported, she made a successful jump.

Flight also reported on an air race she took part in at Orley, an event in which a number of manoeuvres and procedures had to be performed against the clock. Mary was the only woman competitor. In a landing test she damaged her plane slightly but, realising she would not be penalised points if she could replace the part without going over her allotted time on the ground, she decided to make a run for the hangar at the opposite end of the aerodrome. Her athletics training stood her in good stead; she set off at a sprint and kept it up, not only to the hangar, but back too, lugging the replacement part along with her. A reporter present noted that in his opinion not one of the male pilots present could have even approached her time for the dash.

At the beginning of 1927, Major Elliot-Lynn appeared out of the mists of Kenya and, always a chap for the surprise move, introduced Mary to the woman he wished to make his third wife. He was now the one who wanted a divorce, and she was happy to give it to him, since there was a new and younger (by two years) man in her life: Sir James Heath, a 74-year-old ironmaster and colliery owner from Stoke-on-Trent. It wasn't long before the aviatrix and the aristocrat were close enough to consider marriage, which they formalised at Christ Church, Mayfair. The bride wore black. The best man was Mr Stammers, a solicitor who was also an amateur pilot of some distinction. Only a very few close friends were present.

Given Mary's later comment, 'All my life I have struggled for happiness, a husband and a home, but my various romances, none of which were my fault, have ruined me,' one is tempted to wonder where her good sense went, unless, of course, she was counting on the quick demise of said husbands, though one feels that the globetrotting adventures of Major Elliot-Lynn might have given her warning that not all 77- or 74-year-olds are quite ready to sign out.

For at least the first year (since the marriage only lasted three in total), the pair were happy enough; Sir James occupied himself with mastering more iron, presumably, and Mary went back into the air, where she soon began to make a name for herself.

She already held the British altitude record, flying her own Avro Avian with an ADC Cirrus engine to a height of 16 000 feet – as to how she had been able to afford her plane, whether there had been a settlement with the Major or some help from the ironmaster, we don't know. Her companion on this flight was Lady Mary Bailey, wife of the South African industrialist and patriot, Sir Abe Bailey. Lady Bailey had started life in Ireland as Mary Westernra, a fortuitous name in the annals of fictional flying since fellow Irishman Bram Stoker had borrowed it for his bride of Dracula, bat-lady Lucy Westernra. Young Mary had no bitingly bad habits, though she had run away from school in Ascot, England, when she was 12, proving that, when she set her mind to something, she would carry it through, whatever the consequences. When she met Mary Heath, she was equitable, calm, country-loving and married to a man 26 years older than she was (surely there must have been *some* young fellows left after the war?). She had been informed of the nature of the beast only the night before her marriage and though she had never quite come to terms with reproduction, she had six children, was possessed of a lively curiosity and bags of courage and had set her mind upon learning to fly as a way of 'getting away from the pram'.

The two women met at the London Aeroplane Club and, perhaps thinking of picking up a few tips for her own coming elevation to the aristocracy, Mary offered to take Lady Bailey up for a spin. Like Mary before her, Lady Bailey was captivated by the whole experience and signed up for a course of lessons from the club's professional teacher as well as keeping up her co-piloting. She got her licence in October, the same month Mary got her baronet, though there was no invitation to the wedding for Lady Bailey. Perhaps this was just as well; after all, Lady Heath, as she now was, wasn't exactly the real thing.

Lady Bailey began competing in air races and looking to make a few records of her own. She attended an air show in Norwich, where she was characterised as 'camera shy'. This wasn't an unfair description: she was never a woman for the headlines and in her record-breaking attempts was more concerned with beating herself and the conditions than with public applause. Flying had changed her in one respect, though: shortly after gaining her licence she was spinning the propeller of her plane by hand to get it started, when she stumbled just as the engine caught and was hit across the scalp by one end of the prop blade. The aircraft maker Geoffrey de Havilland was present and maintained that he picked up a large piece of skin and hair and handed it back to her. In her biography of Lady Bailey, *Throttle Full Open*, Jane Falloon writes, 'In spite of the gruesome injury, Mary appeared at Bournemouth for a meeting later the same month with her head bound in [a] turban, which she went on wearing for the rest of the year. . . . Her children . . . believed that so great a blow to her head might have contributed to a deterioration of her temper and her inability to control her outbursts – which sadly became the characteristics that anyone who knew her well remembers. The accident, however, did nothing to discourage her from flying.'[1]

At the Bournemouth meeting there was a women's air race with Ladies Heath and Bailey and a Miss O'Brien competing. All three pilots handled their machines well, according to *Flight*, but Mary Heath threw hers round the course markers with particular gusto, virtually standing the craft on one wing-tip. Not surprisingly, she won the race. Maybe she was beginning to feel a contender for Top British Woman Pilot coming up behind her. She wouldn't have been pleased –

---

[1] *Throttle Full Open: A Life of Lady Bailey, Irish Aviatrix* by Jane Falloon, The Lilliput Press, 1999.

she took particular care to present herself as a glamorous aviatrix, often flying wearing high fashion and never, ever looking like a grease monkey. If necessary, she would change in the air before landing: she knew very well the value of appearance, of front, since it was something she'd been projecting, in one way or the other, all her life.

However, at times the strain of keeping up appearances, giving a performance as well as coping with the technical demands of flying, began to make itself felt and she started to find herself drinking more; never *too* much, not yet, just an extra glass to help her keep it all up in the air at the end of a long, tiring day.

In 1927 Lady Bailey broke Mary Heath's altitude record, climbing to a height of 17 283 feet in a de Havilland Moth, a record she kept despite Mary's attempt to take it from her in October 1927, when she reached the same height but failed to advance it by the necessary margin. In July of that year she performed a tour of British aerodromes, visiting over 50 and landing at 33. There were two good reasons for what might seem a rather pointless accomplishment. First, Britain was woefully badly provided with airstrips and no coherent map of these had been constructed; there was also a need to link in the temporary landing grounds that were not up to commercial or club standards but which, in an emergency, might prove to be life-savers. Mary's flight provided the first such plan of the country's landing zones. The second, and equally important, point was publicity. Between 1925 and 1935, the golden age of amateur flight, the public was hungry for sensation, for records and very much as explorers today have their antennae finely tuned for any unexplored spot they can get to with a TV crew, so the flyers of the 1920s and 1930s realised that public support, and it was often a paying public, was keeping them in the business. And then, for Mary Heath, there was always the satisfaction of beating Lady Bailey, which she achieved in the 1927 Ladies Cup Race by a last-minute dive, seen by some as rather unmannerly and oikish, streaking over the winning line yards ahead of her rival.

Somehow, though, for all the silk dresses, the articles penned about every flight she took, the very real achievements not only on her own behalf but for women flyers and athletes everywhere, Mary Heath could not gain public affection to anything like the extent of the quiet, shy (though privately foul-tempered), modest and unhectoring Lady Bailey. In October, she was invited to become President of the Suffolk and Eastern Counties Light Aeroplane Club, whereas Mary had only ever been president of societies she'd invented herself.

Nineteen-twenty-eight began with calls, in a magazine owned by Lord Bailey, for his wife to be awarded the title of Champion Woman Aviator of the World (somehow it doesn't sound quite serious, but it was) ahead of American Ruth Nichols. Then came the news that Lady Bailey was about to set out on a long-distance flight from London to Cape Town, though she stated modestly, this was not an attempt at yet another record, just a bit of trail-blazing and the chance to visit her husband who happened to be at the Cape.

This time Mary Heath was going to be first, in a back-to-front sort of way: she was already in the Cape with Lord Heath, visiting some of his business interests, and decided to make a Cape Town to London flight. The visit had been a real success for Mary and for South African women's aviation: she'd flown in a number of fund-raising events as well as giving free flights and lectures. One of these, 'British Flight', given with lantern slides at the Carlton Hotel, Johannesburg, attracted an enthusiastic audience of 300. A public flying meeting drew thousands of eager spectators to see the races, the balloon catching event (a sport sadly out of fashion nowadays) and the aerobatics. Mary was the first British woman to loop the loop: 'it is merely turning a corner but doing it in a different direction from the corners we are always turning on the ground, and because of the centrifugal force that keeps you in your seat you simply do not realise you are going round till you touch the wake of air you left behind you when you were going into the corner.'

One of the male pilots who flew against Mary at the Cape races was Lieutenant Dick Bentley, the trail-blazer of the London–Cape Town air route. He had set off the year before after having his plane, *Dorys*, christened by Lady Bailey. He had married his fiancée Dorys in South Africa and the couple were about to fly back to Britain; although the Lieutenant gallantly allowed Lady Heath to take off on her Cape–London flight first. He also sold her his maps, the only ones in existence, then asked for them back to check his own route, then lost them!

Mary probably needed all the runway she could get to leave the ground. Her Avro Avian was loaded with everything she might need to use or refit on the trip or if she came down: one spare wheel, with tyre; one set of tools; two spare petrol gauge glasses; one valve with springs; rocker arm; push-rod and nuts; two sets of induction gaskets; asbestos; two sets of cylinder head gaskets; one set of undercarriage-to-mainspar fittings; one pump; one funnel; one chamois; one cleaning cloth; one bottle of dope; string; thread; linen; needles; one bottle of shellac; assorted screws; split pins; bolts; nuts; 10 yards of mosquito netting; one shotgun and 50 rounds of ammunition; one change of underclothing; one pair of mosquito boots; two washing silk dresses; one white flannel skirt; one evening dress; one pair of tennis shoes; one tennis racket; one pair of black satin shoes; six pairs of silk stockings; ordinary toilet-set requisites; one Bible; two blouses; one jersey; one fur coat; one hat; one flying topee; one pair of goggles (tinted and plain); one water bottle; one packet of chocolate, maps, passport, journey logbook, writing materials, reading materials, camera, iodine, lint, bandages, vaseline, morphia and quinine. Total weight: 112 lbs.

She took off from Cape Town on 5 January 1928. The first day's flying was easy but Mary was well aware that things could go wrong at any time. The second leg, from Pretoria to Bulawayo, was about 400 miles. She set off loaded with 42 gallons of petrol – an extra weight that made the controls so sluggish that the plane was hard put to get off the ground but, by using every foot of the airstrip, Mary managed to coax it into the air – bets had been made by a number of local male pilots that she wouldn't – and head off into the baking blue sky. She had packed a flying topee against the sun but was not wearing it that morning; she had only a leather flying helmet which gave no protection from the heat beating down hour after hour as she headed north. She had also packed a number of popular novels to pass the time but they were somewhere in a locker and out of reach. She had no choice but to sit looking down at the veldt unwinding beneath her, mile after mile, hour after hour, a hypnotic progression of time and landscape relieved only by the sinuous windings of the great river Limpopo.

Shortly after noon she was crossing the quartz glare of the Matapas mountain range when her engine started misfiring. She began to look around for potential landing sites. There was nothing on the slopes below but far ahead lay the plains that surrounded Bulawayo and she knew that as long as she could keep the craft in the air, she'd be able to land there in perfect safety. Then she felt a growing pain in her neck and an ache that spread over her shoulders and the back of her head. She had assumed that because the air around her was cold, the effect of the sun would be negligible; she was wrong, and was now experiencing the beginning of sunstroke. She realised that the last time she'd got it, she'd collapsed almost in an instant – not something you'd want to do in an aeroplane at 10 000 feet above a mountain range. She pulled off her slip and wrapped it round her head but, as she wrote, 'the mischief was done. The pains in my back got worse, and black bobs began to dance before me.' She gritted her teeth and tried to control the black spots as they rose up and up and threatened to cut off her vision. She knew she wasn't going to make the plain ahead and, squinting through the dancing shapes filling the windscreen, managed to make out what she hoped was an area of flat land to the north-east. She began to turn towards it. She recalled later that her mind was unusually clear at this moment, she could see the situation and herself in it and then – nothing . . .

... until she opened her eyes to see three native girls sitting around her, smiling. She was lying on the veldt – they had taken off her fur coat and laid it underneath her and were dipping her handkerchiefs in milk and laying them over her forehead, which ached intolerably. Pulling herself up, she blessed her good luck in being thrown clear of the crash but when she looked around for the wreckage, she saw the plane parked nearby, undamaged apart from a drooping wing. The girls helped her to the craft where she checked the time and found she'd been unconscious for over four hours. The plane, which she concluded had virtually landed itself, was flyable but she was in no condition to take the controls: her head was pounding, her vision was seriously fractured and she was violently sick from the exertion of getting up. More than anything she wanted to lie down in the shade and sleep, but she knew she could not relax until the aircraft was secured for the night. With the help of the girls, she collected stones which were bagged to create land anchors to which the plane could be tied in case of high winds. The girls then took her back to their village, which was nearby, and one of them told her that, on landing, she had asked for milk and written a note which she'd requested be delivered to Bulawayo. She remembered nothing of this and a few days later she saw the note which was total gibberish and quite unreadable.

She slept for some hours and then woke to find herself surrounded by the girls' family: she was an honoured, if somewhat delirious and sun-struck, guest and was given food and drink and then allowed to sleep again. In the morning the ache in her shoulders and her head was somewhat better but she was still dizzy and sick. Her hosts had got in touch with a local farmer and she arrived during the morning by car and took Mary, 'weeping like a kid, I don't know why', back to the homestead where, once more, she was put to bed. While she slept, more messages were sent and a Rhodesian Air Force officer came and collected her plane. He was a pilot with thousands of hours of air-time behind him and told Mary a few days later that he'd found the take-off, over bumpy ground and through a maze of trees, almost impossible. Perhaps only the fact that a woman had got the craft down without damage drove him into getting it up again. It brought home to Mary the extreme luck she'd had in landing in one piece, but it did not dampen her appetite for adventure one iota. She thought Lord Rosebery's advice to his son: 'my boy, live dangerously', the finest thing that could be said to any young person. 'Fear is a tonic,' she said, 'and danger should be something of a stimulant.'

By 28 February Mary was quite recovered from sunstroke and she set off again at 5.30 a.m., flying over Victoria Falls and landing in Livingstone, still in the cool of the morning, at 9.15. Here she met up with Lieutenant Bentley, her old flying rival from South Africa, who was travelling back to England with his new wife in leisurely stages and had caught up with and overtaken Mary during her stopover. The two flyers decided to continue together for a few days (this would be convenient for Mary when she reached the Sudan, over which women were forbidden to fly alone) and at 4.00 a.m. on 1 March the two planes took off, heading north once more. This time there were no problems and in a couple of easy stages they reached Nairobi where they stayed with the District Commissioner and played tennis and drank beers on the tennis club veranda as the sun went down. The next day Mary was invited on a buffalo hunt by the Commanding Officer of the King's Rifles. The Bentleys and Mary donned the 'shirt and shorts of safari' and set off with hunters and beaters, 'guns on the cock', ready for action.

Setting aside contemporary feelings about hunting and killing animals for sport, this was a moment when Mary, who had to a large extent, and notwithstanding a rich husband or two, made her own way from that poor childhood in Limerick, must have felt she was truly living a fairy-tale. It was 'one of the biggest thrills of my life, when the grass woke into sound about us like waves of the sea breaking, and one heard and smelt and almost felt a big herd stampeding within half a dozen yards in grass seven feet high. The best moment of my life, I think, as far as the poor old earth is concerned.'

Later in the day Mary shot a young bull in the leg; it had to be finished off by a head shot. The party then moved on and eventually caught up with the herd, at which point there was a lot of shooting and a lot of killing, which we shall leave enshrouded in the mists of time.

On 14 March the two planes took off, heading for Cairo. At the staging points along the way, the flyers experienced a high level of frustration: the telegrams they'd sent to keep the authorities informed of their progress had not been delivered and Mary inveighed against the 'terrible state of affairs' that allowed Britain's colonies to become so careless about the delivery of mail and telegrams. The French, Italians and Belgians were all, in her opinion, a great deal more efficient. Navigation, at least, was no problem once the flyers found themselves over the Blue Nile which would, in the end, lead them all the way to their destination.

When they reached Khartoum they had to stop while arrangements were made for Dick Bentley to be ferried back after accompanying Mary over Sudanese territory. During the stopover there was another visitor, Lady Bailey, on her outward flight from London to Cape Town.

Exactly how the two felt about each other is hard to pin down. In her book, *The Sky's the Limit*, Wendy Boase described the two flyers as 'friendly rivals', though making clear that their radically different temperaments set them on vastly different paths. And they probably were, if not friends, at least close enough to fly together in 1926, when Lady Bailey was learning her craft. However, once she began to claim her own records and to make her own mark in the flying world, as a real aristocrat rather than the bought-in variety, she would have found her path considerably smoothed by family contacts. Mary must have begun to feel a touch of annoyance, even of envy, at this. After all, Mary Heath was no saint. She had a mind and she was not afraid of speaking it, often delighted in doing so and sometimes, as her fondness for taking a drink increased, did so in situations where common sense might have cautioned silence. Lady Bailey, although modest about her achievements, was not given to holding her own temper when frustrated or annoyed. She was also able to afford the best logistic and mechanical backup. Commenting on the meeting in Khartoum, Mary said how very much she admired Lady Bailey's pluck in making her flight, especially since she was doing it without the meticulous care and attention that her mechanics usually lavished on her engine. In the same passage there are so many 'gallants' and 'pluckys' that even the most generous interpreter might begin to suspect a touch of irony. And finally, there was the dinner given at Khartoum in honour of the two aviatrixes. Mary appeared glamorous and young in evening dress; Lady Bailey, looking her age, was forced to wear the tweed suit she'd flown in. Bailey biographer Jane Falloon says that her 'aristocratic background would have carried her through a social situation that might have been a disaster for someone else. She would have had too much innate social confidence to worry about whether Mary Heath was scoring points over her or not.'[2] Well, she might not have worried but I very much doubt if she ever forgave Mary Heath for that evening.

But Falloon does have a point: Mary Bailey's world was built on certainties of class and wealth that were indestructible; Mary Heath's world was not – it was an invention of her own, a very brilliant and beguiling one but, like any fiction, forever vulnerable to the listener who suddenly shouts out that it's not true at all, it's just a story.

In Lieutenant Bentley's company, the two Ladies passed on their way. Mary followed the Nile up to Cairo where she slipped on silk stockings before landing since the city was, after all, the greatest in Africa. After landing she immediately attended to her sponsorship obligations and sent telegrams back to the manufacturers of her engines and spark plugs, praising their performance.

[2] *Throttle Full Open: A Life of Lady Bailey, Irish Aviatrix* by Jane Falloon, The Lilliput Press, 1999.

She relaxed for a few days in Cairo while trying to arrange for the next stages of her flight. Mary's 'gallant little friend' Lady Bailey had found her progress easy, since 'all along the line her approach had been heralded by letters from the Colonial Office'. When Mary asked the local Air Force Commander if *he* could arrange a sea-plane escort across the Mediterranean, the request was brusquely turned down. She was reduced to sending a telegram addressed to Mussolini, Italy, asking if the Italian Air Force could help. Il Duce was tickled by the request and cabled back: 'I have put a sea-plane at your disposal.'

Before the sea, she had to reach the coast, which meant a couple of hops, first to Benghazi, where she found herself 'walking into another great room full of men, the local Italian Air Force', all of whom, after a stiff vermouth, she found to be quite charming; then on to the tiny landing strip at Aghaila, where more charming and charmed Italian officers and another heavy night left her with a nasty hangover. Once she was in the air next day it vanished like magic and she headed along the coast to Tripoli. Here she was fêted once more by the Italian Air Force, which seems, *en masse*, to have taken her to its heart. Her youth, her *bella figura*, her courage and her style must have seemed like a visiting dream to these men trapped in their desert postings. Mary thought they would have liked to keep her in Tripoli forever but they couldn't and she flew away, over the soft green and deep blue of the Mediterranean.

Once over Europe, Mary began to feel she was almost home. She wasn't; nothing is ever sure in distance flying except that something will go wrong when you least expect it, and probably when you most expect it too, and also when you feel neither one way or the other. Something always goes wrong. Over Marseilles the weather closed in and after flying thousands of miles through clear skies, she found herself in thick cloud. Rapidly descending to 60 feet, she followed the railway line toward Dijon, landing there more tired after the non-stop concentration than after days of flying over Africa. From Dijon to Paris was, if anything, worse: two storms blew up and her Avian was thrown about in the sky and she was bounced around the cockpit until she was one vast bruise and she feared the wings would be snatched off the fuselage. She landed safely in Paris and spent the night there before setting off on the last stage, across the Channel, in another storm, landing at Lympne in Kent before making a last hop to Croydon. Over the airfield she was joined by two escorts and, looking down, she saw a huge crowd waiting to meet her. She couldn't resist doing a loop of triumph before coming into land, something that her little friend would never have contemplated.

The crowd surged forward as she stepped from the cockpit, perfectly turned out as always, in a black straw hat and walking dress, and accepted a bouquet of roses. There was a large contingent from her engine builders and after they raised a cheer for her, she raised one for them and the engine which had performed so well over the 10 000-mile flight. Lord Heath was there to welcome his wife back and can be seen smiling under his moustache, slightly to the edge of the picture in one of the official photographs.

A week later a dinner was given in Lady Mary Heath's honour at the Mayfair Hotel by the Air League of the British Empire, in association with the Royal Aeronautical Society, the Royal Aero Club and the Society of British Aircraft Constructors. His Grace the Duke of Sutherland presided. Under-Secretary for Air, Sir Sefton Brancker was present, as were many more of the great and good in the world of flying. Speeches were made praising the achievement – the first woman to fly alone from Cape Town to London – mentioning her past as athlete and lecturer at Aberdeen University (perhaps it's best not to look too closely there) and stating that the air age was definitely here to stay. One speaker mentioned a moment when, flying over Nairobi, turbulence had prompted Lady Heath to lighten the plane by throwing out her tennis racket and a number of the novels she'd brought along to pass the time; he wondered which ones she'd thrown out and which she'd kept. She didn't say, but she did make a gracious speech of thanks, mentioning Lieutenant Bentley's invaluable assistance, as well as the many other people who had supported her on the journey. She finished with the suggestion that the firms connected with the flight might like to subscribe to a fund setting up four flying scholarships, since it was clear now that the whole country was dying to get into touch with aviation.

It was a great evening, the height of her career; things would never be better.

Her marriage to Sir James, while able to sustain itself amid the blur of public events was not, and never had been particularly well founded. Mary got a social position, albeit one which could be jerked away at a moment's notice; money enough to buy three planes and travel; contacts in the world of politics and business. Sir James' side of the bargain is unknown: was Mary a trophy wife; was there affection and/or lust between them? In July 1928 they gave an air garden party at Croydon but otherwise seem to have appeared very little in public together. In August, Mary announced that she would be using her B licence as a commercial pilot for the Dutch airline KLM. Does this mean she was running out of money or that she was pushing at the barriers again – KLM's archives have no note of her either applying or being accepted as a pilot for the company, so perhaps the announcement was intended to prompt some action by her husband. We simply don't know.

Aside from the social aspects of her life, there was the constant pressure to conform. At a meeting in Orley later in the year, a commentator wrote of the 'inevitable note of comedy introduced into proceedings by Lady Heath'. Then there were unconfirmed rumours of a faked attempt at the height record, though, since she got this back from Lady Bailey at the end of the year there seems no point in lying about it. Unless there was an imp of the perverse at work somewhere, upsetting arrangements and spoiling every triumph, playing pranks for the fun of it. More unconfirmed rumours hint at an unseemly scene when Mary met the American humorist and flying enthusiast Will Rogers in

London. It is said she was drunk and vulgar; she may just have been loud and holding a martini, or the whole thing may be no more than a fabrication. British flyer John Courtney, who'd delivered newspapers with her by air during the General Strike, said that after she became Lady Heath she turned into an alcoholic nuisance – but one woman's cocktail may well be another man's drunk and Elinor Smith, who knew both of them, had harsh words for what she called Courtney's casual dismissal of Lady Heath.

Back in South Africa, Mary's plucky little friend Lady Bailey had arrived for a triumph of her own. Sir Abe Bailey was a major figure in South African political life so it isn't surprising that his wife got what was virtually a State reception with film cameras, reporters and a vast crowd. Proving that she had learnt something from Mary Heath, she climbed from the cockpit to reveal a smart woollen outfit with a large felt hat worn over her flying helmet. She too was honoured with dinners and interviews and there can be no doubt that her flight, though eased by a far more efficient and powerful backup organisation, was as much a feat as Lady Heath's. And then she announced that she intended to top it by flying all the way back to London, thus making the first two-way journey between the cities. First, though, she and Sir Abe would spend some time together on one of the family's extensive properties.

Rest was the last thing on Mary Heath's mind that autumn. In September she was touring Ireland giving demonstrations and lectures; she was also part of the British team at the Orley air show; in October she regained the light aircraft height record, reaching 23 000 feet. No oxygen was used; she took 1 hour 10 minutes to get up there and 10 minutes to come back down. She also managed a height record in an all-metal sea-plane and a second appearance with the British team at Rotterdam.

The October 1928 issue of *Flight* announced that Lady Heath would be making an air tour of America to exhibit her de Havilland Moth and act as test pilot for various US manufacturers. The trip would begin just weeks before Lady

Bailey arrived back in Britain at the conclusion of her two-way flight, which was already being called the greatest solo effort ever accomplished by a woman. Mary probably wasn't too down-hearted at putting an ocean between herself and her plucky little rival.

Mary's first official engagement in America was at the Chicago International Aeronautical Exposition, the largest air show yet mounted, where new planes and engines were being exhibited to an increasingly air-minded public. Lady Heath was demonstrating both her de Havilland Moth with its 'gypsy' engine and the Avro Avian (borrowed back from Amelia Earhart, who had bought it) in which she had made her record flight. She was also acting as the US correspondent for *Flight*, sending back a series of concise reports on the latest developments in air technology.

On a nearby stand was the ex-World War fighter pilot and England–Australia flyer Bill Lancaster, demonstrating the Cirrus engine. He was living in America with his mistress Chubbie Miller, who had flown with him to her native Australia, but was in Chicago with his wife Kiki, who had paid him an unwelcome surprise visit. It was an interesting and difficult situation for husband and wife, who weren't exactly estranged but weren't exactly together either. Kiki was missing her children and Bill was missing Chubbie. Mary, never one to judge the behaviour of others, became friendly with both Bill and his wife. So much so that when, after the show, she decided to fly down to Miami to compete in the air races there, she invited Kiki, who had never flown, to accompany her on the trip. This got Bill and his wife out of a tricky situation: it also gave Kiki Lancaster the chance to shine a little herself, rather than always being the figure in the background, mentioned in news articles as the loyal helpmeet (Bill's relationship with Chubbie had yet to hit the front pages in a national scandal).

The flight attracted a certain amount of press attention; women in the air, at least, British aristocrats in the air, were still something of a novelty, and there were enough adventures and alarms during the trip to keep readers interested, including an irate farmer in Savannah who was going to shoot them for crushing his crops and a beach landing at Daytona with the sand hills inches away on one side and the sea coming in on the other. Kiki loved the trip and went back to England, if not happy with her errant husband, at least content to have tried her wings at last.

At the Miami races, Mary won two events and began to consider the possibility of a life in America. It seems clear that she had made no plans to return to Britain and her husband; there were press reports that she intended to travel on to Australia or that she was thinking of applying for American citizenship: the marriage was, to all extents and purposes, over, though nothing had been said and, unless something came up to alter the situation radically, nothing would be. Mary was still receiving a small allowance from Lord Heath but was, both temperamentally and of necessity, earning her own living.

After the Miami meeting, Mary demonstrated her plane and its engine round the country, covering 3000 miles and gaining more than a hundred orders for the Cirrus engine. She also continued to send back entertaining and well-informed reports on the new planes she tested:

> The third machine I tried was, to my mind, the pleasantest: the Huskey Junior made by the Consolidated Air Craft Co. of Buffalo. With a 110 hp engine, this aircraft lands under 40 mph. It had no air-speed indicator but I reckon she did not drop until about 32 mph . . . I felt it was a privilege to fly this machine and hated bringing her down. The controls are beautifully balanced and she stunts as if she wanted to. One was even tempted to try her upside down a little way, a performance she takes very well owing to the wing section . . . One has a curious feeling in this aeroplane that it would be utterly impossible under any circumstances to break it.[3]

[3] Testing American Aircraft, by Lady Heath. *Flight*, 17 January 1929.

In February she was protesting to the US Department of Commerce because they would not allow her to take the test for the American B licence, which would allow her to fly commercially in the States. She always wrote a good letter, especially when her ire was aroused in a cause she believed in – whether it was getting women the right to fly commercial aircraft or earning a living for herself – and she was granted that right in a short time. Ironically, she failed her first test but passed a second and went on to register a company in New York to handle and promote herself as a flyer and personality. Known as Aerial Activities, it was, perhaps, ahead of its time; it certainly came into conflict with G. P. Putnam, who was then marketing his wife Amelia Earhart with a degree of expertise that would make her one of the most famous women in the world. Mary Heath recognised the power of publicity, and had done so ever since her story-telling days at school in Limerick, and was happy to share her plans with Putnam, little realising that he would use this knowledge to undercut and oust her from profitable lecture tours in favour of Earhart. That was the future; at the time she saw herself as living in a society where advertising and the publicity machine were considered a natural part of life rather than as, in Britain, something somewhat distasteful and loud (like Lady Hell-of-a-din) about which one would rather not talk. Had things gone well, had Putnam not sabotaged her career both in the air and on the lecture platform, there's no doubt that Mary Heath would have become a national, even an international figure.

In mid-summer 1929 she was invited to take part in the first women's transcontinental race, the Powder Puff Derby, but decided instead to enter the Cleveland Air Races held at the conclusion of the Derby. She had met many of the competitors as well as other American flyers, including Marvel Crossen, who was to die during the race, and Elinor Smith, another pilot who realised the value of publicity and had received a great deal of it when, still a teenager, she flew under the bridges of Manhattan's East River. Mary and Elinor were to compare notes about their meetings with G. P. Putnam, who had offered his services to Elinor – and Bill Lancaster – and had treated them all in a similar manner. Neither woman held it against Earhart herself, though it couldn't be said that either of them actually liked Lady Lindy and, for Mary, who had offered friendship and sold Earhart her record-breaking Avro Avian when she was in England after her first Atlantic flight as passenger, it must have seemed like a cold return indeed. Mary and Elinor, both on the feisty end of the spectrum, got on well and made a number of flights together, Mary offering the teenager advice about dress and presentation, telling her that there was never an excuse for looking like a grease monkey in overalls; that a Lady always managed to climb from the cockpit dressed appropriately.

In late August, Mary was offered the chance to test-fly a new American ship. It was a run-of-the-mill job but with lecturing and exhibition flying drying up, it would bring in a few dollars and take no more than an hour or two; but things go wrong, and on this occasion, they went catastrophically so. Flying low over an industrial area, she was forced to make a sudden turn to avoid a factory chimney that was taller than she'd estimated. Pulling the craft round, she caused the engine to stall and her plane cartwheeled on to the glass roof of a factory, smashing through it and plummeting down onto the factory floor. Fortunately no employees were hurt and they were able, after a considerable length of time, to cut her out of the twisted metal of the cockpit. She had internal injuries, broken bones and a fractured skull. She was not expected to live.

She underwent extensive surgery and, helped by her naturally good constitution, began to recover. It was a slow process. By mid-September she was able to get out of bed and walk around the hospital and start a course of physiotherapy. This brought immediate results and she was discharged in October. Reports in the British press speculated on whether Lady Heath would ever fly again. She had no doubts: a few months of rest would see her fit and ready to attempt yet more records. She moved to Reno, Nevada, where the heat would aid the aches of mending bones, and began to plan for the future. Things were looking up, she thought: she'd just met a man who might, for the first time, provide the love, companionship and home she said that she'd always craved.

Back in Britain the fame she'd always craved was settling about the modest shoulders of Lady Bailey. After her return to Croydon in January, where a slightly bigger crowd than had greeted Mary was waiting, she was given a celebratory dinner with a slightly larger and more glittering assemblage of guests at the slightly grander Savoy Hotel. Sir Hugh Trenchard made a speech, the Minister for Air seconded it, Lady Bailey replied modestly and everybody loved her; it was the sort of throwaway self-denigration reflecting an upbringing that taught you never, ever to trumpet your own virtues. It was very British and very attractive and, for all her skill in story-telling, for all her passionate belief in herself and her causes, Mary Heath never really understood that the best way to get anyone to believe you is to let them do 90 per cent of the work themselves. In March, Lady Bailey was elected Chairman of the Ladies Committee of the Air League, and after that, a whole raft of honours and citations and awards was waiting to be offered and graciously, modestly, accepted.

Mary Heath's only official appearance in London was made in absentia when she was sued by a firm of dressmakers for unpaid bills. One of the principal prosecution witnesses in the case was the part-time pilot Mr Stammers, Sir James Heath's solicitor and the best man at his marriage to Mary. This might lead to the conclusion that Sir James had definitely stopped picking up the tab for his wife. And perhaps one can understand his feelings, since news came that Mary had divorced him in Reno, an ineffective divorce under British law, and had married the airman William Williams, who called himself, not unreasonably, George, and was approximately the same age as his bride.

The marriage took place in Longville, Kentucky with local Mayor, Flem W. Sampson officiating, until it was discovered that he didn't have the proper authority and the local priest was rousted out to do the job – not that the marriage was legal, since Mary was still, in British eyes, married to Sir James, who was in no hurry to exercise his lordly clemency and oblige her with anything at all. He commented, 'I was served with the usual legal papers but I have taken no notice. I have heard nothing about it since, except what I have read in the newspapers.'

One spot of good news for the couple was announced in *Flight* magazine, when it reported that Mary had been awarded $3000 damages for her flying accident by the Workmen's Compensation Board. However, the records of the Board show no such award as having been made, and indeed the accident was her fault, so the whole story seems unlikely. Possibly it was planted by Mary and George, who were getting down to their 'last Mexican ha'penny' to throw creditors off their track as they started on an aerial honeymoon. They visited Italy and points east, gaining a certain notoriety – vitriol to Sir James's tender vanity, one hopes – when they offered their services as fighter pilots to the Chinese Air Force in its operations against the Japanese. Once more, Mary's heart was in exactly the right place with the right cause, but their offer was turned down and they flew on.

In 1932, the 80-year-old Sir James felt the first stirrings of a new love – he was to marry wife number four, Dorothy Mary Hodgson BSc in 1935 – on the horizon since he issued divorce papers against Mary, obviously the manly way to do things, naming her adultery with Williams and the ceremony of marriage in Kentucky as good cause. The suit wasn't defended and, on their return to Ireland, Mary and George married once more, legally this time.

There remained the problem of making a living. From all we know about George Williams, and it isn't a lot, he wasn't particularly good at that side of things. He was around when Mary started a flying school in Ireland but seems to have made little impression. As did the school itself, which failed to make a profit and was sold. Williams disappears from the records and Mary begins to sink out of sight herself – in 1933 the minutes of the Irish Air Club report a quarrel: 'Mrs Williams states in the firmest possible way that if she is not given free lifetime membership of the club in recognition of the flying scholarships she has raised, she will have nothing more to do with the institution.' The minutes note that the matter will be dealt with by letter.

Throughout the 1930s her drinking gets worse. It is quite possible that she was in some degree of pain from the 1929 air crash. She was certainly chronically short of money; she had never been one to save and the four planes she had once owned were long sold by now, and this resulted in her being excluded from the flying world that meant so much to her.

She was, in fact, slipping into that social limbo which caused the middle classes of the 1920s and 1930s to sweat with sheer terror. There was no safety net, no social services to lend a helping hand – if you fell, you fell all the way; you became declassed and, rather than sympathy or help, attracted blame, as if your social state reflected your spiritual condition. And in Mary's fall, her acquaintances must have seen more than a touch of *schadenfreude*; she'd been playing fast and loose with respectable society for many years, blowing her own trumpet, cocking a snook to many and was now receiving her just desserts. And for those friends who genuinely wanted to help – and they seem to have been remarkably thin on the ground – her continual drinking, bad language, shouting, endless promises to reform and then her slide back into drink must eventually have become too much. Maybe Mary herself decided to drop out of sight; drunk or not, she still had her pride and her achievements and her hopes.

In 1936, Mary Williams was back in London and living in a hotel near Paddington Station. She was still drinking heavily to dull the pain and the loss, and trying to remake old contacts. One afternoon she was picked up drunk and disorderly in public. She was sobered up in the cells and called before the Magistrates at Bow Street Court, where she was ordered to provide a surety for her own future sobriety. She had no money, so she couldn't provide such a surety. In a city where a million men got drunk every Friday night, one drunk woman was committed to prison for 40 days. At least somebody in the judicial system had a shred of common-sense left and Mary spent only two nights in Holloway. Of the experience she said, 'It was the first time in my life that I have been really happy. I loved being among the women, they were so kind. I wish they had left me there.' There is undoubtedly some truth in the words. Mary had been inventing herself all her life, writing her own fiction and playing word-tricks on the world around her. In prison she didn't have to tell any more stories; for a couple of nights at least, she was at rest and, maybe, at peace before being released on bail in the custody of friends.

Within hours she was missing again, and stayed that way for four days, finally being located unconscious in a Birmingham street with three pence in her pocket.

She did make one more public appearance, in a newspaper interview where she talked a little about the glory days when she earned £3000 a year as a flyer. She also shared her hopes of returning to Ireland and taking up her medical studies again. She said she had stopped drinking and had been to church last Sunday and was looking forward to a new start.

Three years later she fell from the tram and fractured her skull for a second time. After lying in hospital for two days, someone went through her bag and found the name Williams. Eventually a relative was traced who identified the unknown woman. She was cremated at Golders Green Crematorium on 9 May 1939. She had been living in one room at Cooks Hotel, Paddington. Her estate amounted to £240 7s 10d.

5
Evelyn bobs her hair

"...it's a long time, looking back all that time. Some things I guess, I just forget 'em. There's so much, so many friends and good friends and ... good times and bad times too. But you get over things. Oh my! You do, you just have to. You just have to get on with life. But all the same, when you try to remember ... Well, I remember my mother, she would, before I went to school ..."

I n the mornings, waiting to go to school, she would sit in the kitchen while her Mom cooked egg and bacon for breakfast and cut thick slices of bread for sandwiches. Mom would ask her what she wanted but it didn't really make a lot of difference because it was usually jam and fruit, always a piece of fruit in the brown paper bag.

Sometimes Mom would give her some money and send her out to the bakers to get a warm loaf 'and enough boiled ham for all of us, honey', and she would make what Evelyn thought were pretty good ham sandwiches.

When she went to school she wore a coat and blouse and a skirt and long socks and lace-up shoes like all the other girls but she thought she'd rather wear long trousers like the boys because it seemed good sense if you wanted to run and jump and climb trees. Only trouble was, Mom wouldn't buy them and she wouldn't sew them either so Evelyn reckoned she'd just have to wait until her little brother grew and she could borrow his.

Meanwhile, there were lots of things to keep you busy around the place because they moved a lot and the place changed from one year to another. They went from Greenup, Illinois to Oak Creek, Colorado; from Seattle, Washington to Fort Collins, Colorado, because her Dad had these ideas about how to make money and they'd go to a new town and he'd set up a business that was sure to make them all rich. And sometimes it worked but mostly it didn't and then it was down to her Mom. She had some money of her own which her Dad had given her and she ran a chemist and a haberdashery shop in Wyoming, where they all stayed for a while because she was bringing in the money.

But Dad still had his ideas and sometimes he'd go off to start another business and when it was up and running, he said, he'd send for them. But he didn't, not always. There were quarrels that she overheard, between Mom and Dad, about all this moving on for no reason that Mom could see, because he was never going to make their fortunes, they'd be better off just sticking to what they had. But her dad could never see it that way and she sort of agreed with him sometimes, but mostly she didn't.

She had ideas of her own about making money. One of them was about this place that had a lot of old iron in the back yard. She and her brother Denny, who was two years old by now, used to pile it up in a little red truck they had and pull it round to the front of the block where there was a shop that would buy scrap metal. The people there would give them some change for the iron, and maybe an ice cream or a cookie. One day they even bought her a bright orange teddy bear suit which she wore as long as she could – only it wasn't that long, because she was growing all the time.

They used to go into the hills and gather dry wood to make bonfires and cook potatoes in the ashes, flicking them out with forked sticks, trying to hold them without burning their hands – you had to throw from hand to hand all the time, keep them on the move, keep them in the air as much as possible. And since they were out in the foothills and everyone knew there was gold in the mountains, she got all her friends to start digging along the river bank; they made quite a tunnel but then the local judge rode by and told them they'd better watch out, they'd dug so far and so deep City Hall was about to fall in the hole. The only gold she ever found was in her mouth when she lost a tooth and her uncle, who was a dentist, put in a real gold one to replace it. But it was only a milk tooth and soon it had to go because her grown-up teeth wanted to come through.

As time passed and she got taller she began to sort out things for herself. Most of all, she wanted to be a carpenter; she thought that would be grand: hammering nails, cutting wood, building things that would be useful in the house, making toys she and her little brother could play with. So she built a boat. It had a flat bottom and planks for sides and the front and back were flat too. She and Denny dragged it down to the river, which was hard work because Denny was only six then, and she was eight and though she reckoned she was strong for her size it was a heavy boat and the ground was rough around the river bank. But they got it there in the end and slid it down into the muddy brown water. And it floated – there was a bit of water coming in at the sides but she knew that an empty coffee tin would take care of that, if one of them baled and one of them was the captain. They climbed in and cast off and a bit more water seeped through the boards but Denny baled like a sailor and they moved out into mid-stream and started to go along with the current. Luckily it wasn't a very big river and the current was sluggish because they didn't go that far before they hit the bank and climbed out – but the boat was still floating, and she thought that if you *really* wanted to do something, then you could do it. The only thing was, you had to know who to trust, and that was your family and, most of all, yourself.

She was staying with her aunt Irene who kept a small hotel in Seattle. This was in 1916; there was a war in Europe and folk were saying that soon American boys would be going over there too, though a lot of them weren't so sure that the country should be getting mixed up in Old World business like this. Only President Wilson seemed set on it and the Senate was behind him and her cousin was in the Officers Training Corps at college, and the boys all thought that going to France to teach the Germans a lesson would be a fine thing to do – except the ones from German families, who kept quiet about it.

In the evenings her cousin would take Evelyn through the manual of arms until she knew all there was to know about drill and stripping and reassembling and firing her rifle. Except she didn't own a rifle and she wanted to have one almost as much as she wanted to be a soldier. She reckoned there wasn't much chance of that, so she settled on driving an ambulance in France if she ever got there and getting her Uncle Earl to buy her a BB gun. Which, in the end, he did and she started to try all the things she'd practised with her cousin for real. As everyone knew, Germans didn't just sit around waiting to be shot at: she needed moving targets and thought the chickens on a nearby lot would be perfect. They moved alright, like lightning, and she didn't hit but one for hours, and that was an accident, and it wasn't a chicken, it was the neighbour's prize rooster. She hadn't noticed it was different, only that it seemed to be acting in a funny way. But not any more.

She thought there was a chance of getting her hands on a 22-calibre rifle that she'd seen. It was in the window of the pawn shop that was on the ground floor of her aunt's hotel. Whoever pledged it never came back and it was for sale, and cheap too. Then one day it happened that a judge who was lodging in a room on the top floor of the hotel had a problem. He seemed like a nice man and he'd stop and have a word with her, about whatever she was doing, some woodwork or her fishing, which he used to like too. One day he locked himself out of his room. He couldn't get back in without calling the local locksmith, which he didn't want to do on account of how it wouldn't look good, the judge getting shut out of his own room. So he asked Evelyn if she would climb through the little light over the door – it was small but she was slim and limber and might just wriggle through, and then she'd be able to spring the lock on the inside. She didn't fancy it; the space looked too narrow and what if she got stuck there? And the judge told her, 'If you unlock my door, you can have anything you want.'

'Anything at all?' she asked.

'Anything.'

They got a stepladder and she climbed up and started to slide through the gap; she was right, it was narrow and she nearly got stuck but she managed to swing herself down inside the hall and unlock the door. And the judge said, 'So what do you want, Evelyn?' And she said: 'I want that .22 there in the window of the pawn shop.' And he laughed and said, sure, but bought her some sweets instead. She never really believed in the law after that. She used to say, even after she had grown up: 'Never trust a judge'.

At about this time she decided she wanted to take woodwork at school, because she had ideas about becoming a builder one day, or even an architect and designing buildings. She went to see the head teacher and asked if she could stop doing cookery and do woodwork. He said he'd never heard of a girl doing that before and she asked if that meant girls couldn't. He thought about it and said there was no reason he could see why girls shouldn't, so she did woodwork with the boys. Later, when her family moved to San Diego, the new school made her take cooking again and she learned how to make a hamburger.

Before they left Wyoming she was making her own money selling subscriptions to magazines around town; she worked hard and sold a lot of subscriptions because apart from what she earned, there was a special offer to those kids who sold most – you could get a real film projector for yourself. She wanted that projector badly, she loved the cinema in town but most of all she thought about showing her own films in the front room and selling tickets and maybe ice-cream too. And because she worked hard and folk seemed to like her and people in those days bought a lot of things they needed on subscription, she sold enough and won the prize. She used to wake up at night, waiting for it to arrive and worrying that it would be too big to get through the front door, and maybe it would use too much electricity and take up too much room. Then when it did arrive it was tiny, you could almost hold it in one hand, and it had one bit of film with it that she ran right away. There was a bridge, and a boy stood on the parapet and jumped into the water, then if you ran the projector in reverse, the boy jumped right back out of the water and landed on the bridge again. It was weird being able to make him do that: in the water, out of the water. She promised her brother Denny and a couple of his friends that if they helped she would share any money with them, and they set up the projector in the front room and arranged the seats and invited people, it was mostly family, to come and buy a ticket for the film. They raised 50 cents and everyone had a good time but Evelyn went off and bought fish hooks with all the profits, because she liked fishing. Her little brother and his friends were furious – they had other ideas about what to do with their share – and her Mom wasn't pleased either and, in the end, because she'd broken her word, she got into trouble. Which taught her something too.

It began to look like things between her Mom and Dad would never mend and for a while she went to live in Hamilton, Ontario with her Aunt Edna and Uncle William. Then she went back to St Louis to help her Mom at the hat shop she was running – but by now something had happened which was going to change everything. One afternoon she was walking back from school and she heard a noise – maybe she didn't really *hear* it at first, it was just there, right on the edge of her hearing, creeping towards her. At first she thought it was a motorcycle or a lorry, then she realised the sound wasn't coming from down here but from up there. It was a plane. She'd seen a few, seen them in films, in magazines and books; planes weren't that unusual, but then you didn't see them every day either so she looked up, and right there, coming out of the sun, was the most exciting thing she'd ever seen. A plane in the air. Just that. Passing over the road, over the town, its shadow racing beneath it, the perspective changing every moment as she saw the whirling prop and the engine cowling, caught a glimpse of the oval of the cockpit glass and the pilot, then the underside of the wings, the wheels and the fuselage as it flashed over straight above her. She turned to follow it as it flew away, heading for the town which would be laid open beneath it, streets, houses, gardens, everything visible from *up there*. And she was held in a sort of tension between that *up there* and the solid ground on which she stood. It was like holding the string of a kite, being attached to something which escaped all the rules that existed down here on earth and she felt a huge excitement which grew into an utterly confident assertion as she said the words to herself: 'I've got to fly some day.'

First, however, there was other business to take care of. She and her brother got caught up in the influenza epidemic sweeping Europe and North America in the aftermath of the First World War. They were both hit hard and recovery took a long time – it was decided they should go and stay with Evelyn's dad in California, where the warm weather would aid their convalescence. Evelyn wasn't so sure about that;

she had missed Mom while she was in Canada and though she regretted the way her dad got his wild ideas and ran off and she could see how it was difficult for her parents to live together, she still wanted a family that was all in one place at the same time. That was the future, like flying and going to college: right now was the trip to Los Angeles, then on to San Diego and the cookery lessons, then another hop to Chula Vista where Dad was working with the phone company and where they all stayed in her Grandma's house.

She was 15 now and thinking of a career in architecture. She loved the idea of building things that would stand for years and be used by thousands of people; she was also reading about aeroplanes and flyers like Blanche Stewart-Scott, who had been the only woman taught to fly by pioneer Glenn Curtiss. What really interested Evelyn was not only Blanche's determination to succeed, and she took that lesson very much to heart, but the aviatrix's costume. Billed as the Tomboy of the Air, Scott wore trousers; she wore them over two petticoats but she still wore them. Evelyn, who was wearing her little brother's trousers most of the time now (he was getting pretty big, she'd have to take them in soon) reckoned this made good sense. She decided that once she got in the air she'd wear trousers there, just like everywhere else. Except at school of course, where the dreadful cookery lessons went on and on and on . . .

But there was always sports and she was able to put a lot of her energy into the swimming club, competing at meets all over the state and becoming the president of the club. At weekends the club would gather at the beach for practice or just to catch the sun, and one Saturday morning she saw this girl with simply the best haircut: it was close to her head, closer than a bob, more like a boy's cut and perfect for swimming and even better for Evelyn, who headed back into town and went straight to the barber's and told him to cut it all off. It took a while to convince him but, in the end, he did as he was asked. And it wasn't that shocking; girls had been bobbing their hair for a while now and this was just a few snips shorter and closer to the head.

She sat there in the chair, smelling the lotions and the soaps and the garlic on the barber's breath and looking at the gentlemen waiting their turn in the big mirror over the sink and then it was done and she paid and walked out, feeling the air cool around the back of her neck, knowing that this was right and the way she wanted it to be. Mother wasn't so pleased. She'd only moved back from St Louis a few months before, mainly persuaded by Evelyn, who had been missing her and had made the journey back to her old home expressly to convince Mom that they should all be together again in California. And now here was this very same daughter arriving home with a haircut no respectable woman would have been seen in outside of film – the same daughter who habitually wore trousers and thought about studying to be a meteorologist rather than settling down to make a home. There she was, standing there, as bold as brass and saying, 'Well, it's not that short, it's only a bob so why don't you just call me Bobbi from now on?'

Things weren't only chilly around the back of Bobbi's neck for a few days – until she had another of her ideas.

She'd been reading a story in a magazine about two girls who opened a petrol station and made good money, so why didn't they do that too? Bobbi and Mom could run it while Dad kept working for the phone company. Mother, who was going to have to provide the money, told her, 'I should say not.' Dad, who was going to put up the buildings, said, 'Why not?' It was all a bit too like the old family story, another brilliant idea which Mrs Trout was supposed to finance with the money she'd inherited from her father. The only difference, this time, was that the idea came from Bobbi and not her husband, and somehow Bobbi's ideas always seemed to work while her husband's never did. And there was always the problem of his drinking, which didn't seem to be getting any better even if it wasn't getting much worse. And she was a good needlewoman, so maybe she should run up a couple of smart uniforms for Evelyn and herself, skirts and jackets with an initial on the pocket. Why not? It might work, after all.

And it did. Dad put the building up on Elm, they got a contract with the petrol company, they put in an air pump and installed a radio, so customers could sit and listen to a song or a comedy show while they got their tanks filled and their windscreens washed, and that's how it came to be called the Radio Service Station. There was Bobbi on the forecourt, in her smart new uniform, ready with a smile and a helping hand, and Mother in the office in her smart new uniform, ready to take the money. And take money they did, the country was booming and Mr and Mrs America were out on the road driving cars that any family could afford to buy and run on a working man's wage. Bobbi reckoned a lot of folk stopped by just to listen to the radio or have a chat by the pumps, but they also bought the services and it soon became clear the station was too successful and was keeping Bobbi from her education. So Dad left his job and started working full-time on the forecourt while Bobbi went back to school and worked weekends – and managed to lose her uniform somewhere so she was soon back in trousers and shirt – though she always took care to wear lipstick, to keep the customers happy.

One day, while she was filling a customer's tank, Bobbi was talking about school and the swimming club and her ambitions, and she mentioned that she was keen to find out more about flying. This customer, W. E. 'Tommy' Thomas owned a plane, a Curtiss Jenny, which he kept at nearby Rogers Airfield. He offered her the chance of a flight. She accepted. It was 27 December 1922. A late Christmas present, she thought, and went down to the strip where Thomas took her up for a flight. It was like being 12 years old and seeing that plane go by all over again, only this time she was on the other end of the kite string, up there among the birds, slipping and sliding through the air currents, swooping and climbing and falling away *towards* the sun. Now she knew exactly what she wanted to do – she was going to learn to fly.

Back at the service station things were not going so well between her parents. Wanting something a lot, she discovered, didn't always mean it was going to happen and her mother and father were finding it impossible to live together, even with a successful business to offset the tension. She and Mom bought Dad out and he went off to Kentucky to try his hand at . . . any number of new schemes. They kept running the station, though Bobbi registered at the University of Southern California to study architecture, an interest that was rapidly giving way to her desire to get back in the air. She needed money to pay for a course in flying and, being Bobbi, began to push up her part-time hours at the station so she could put aside more cash. The situation changed yet again when her dad came back from Kentucky because he'd lost his money speculating on horses; he said his luck at the track was as bad as it was in commerce; so he moved to Anaheim where he started another garage which Bobbi began to manage for him. Maybe he'd begun to see that his daughter's inventiveness and luck, unlike his, was good.

The new service station did well enough until one of the big companies saw the opportunity and opened an outlet of their own just along the road. Trade began to fall off, oddballs started drifting in: one customer demanded that Bobbi empty the old, used air out of his tyres before she put in the new, free, fresh air. Then her dad fell in love with someone and got married right away with, Bobbi reckoned, about the same chance of success as most of his business ventures; which turned out to be wrong for once, since her dad and his new wife stayed together until she died in the 1950s. The new wife merited, in his eyes, relocation and a new job and so the station was sold. Bobbi's share of the proceeds, added to what she'd saved, gave her $2500. She thought that might be enough to get a lesson or two and hire a plane.

Burdette Fuller was known by everyone as Pop Fuller, and he owned a flying school which operated out of South Eastern Avenue, which happened to be the closest place Bobbi could find, so she went to see him and ask what it would cost to learn to fly. Pop reckoned about $250 to reach her solo flight and, after that, 10 hours in the air to earn her private licence. Once she'd got that, the hire of the plane would be up to her. Bobbi wrote out a cheque, put it on the desk and said, 'So tell me, Pop, what makes an aeroplane fly?' He drew some pictures and explained about uplift and rudders and ailerons and arranged to meet her for her first lesson on 1 January 1928.

She was there bright and early, too early for Pop, so she hung around, kicking her heels, looking at the planes parked outside their hangars, until he arrived and introduced her to one of his Curtiss Jennies, a Glenn Curtiss-designed ship that was noted for its sturdy qualities both in regular flying, he explained, and for any kind of stunting. Some pilots around the strip said those sturdy qualities made the ship fly like an airborne brick but as far as Bobbi was concerned, anything that got her off the ground was going to be an ideal trainer and she took to it immediately. She had no desire to throw it around and see how sturdy it was; right from the start she had a respect for the virtues and vices of the ship that was carrying her up there and never felt that any plane should be wrung out in the air.

Flying felt natural. She soon began to get the hang of the controls and the feel of the plane, and spent as much time as she could around the airfield. She took up a course on meteorology at USC but hated the mathematics and let it slide. For Bobbi, it was always 'hands on', like turning wood or fashioning the gleaming of an idea into something practical. She was not a dreamer; her world existed all around her and was intensely physical, there to be tried and mastered.

By the beginning of March 1928 she felt she was beginning to master her plane. She was also learning about the engine, since that dictated an awful lot of what any pilot could do. As soon as you took off, in that danger time when you were betwixt and between, you looked around for anywhere you could land fast if the motor stopped. You always had to be aware of that, and the better you knew your engine the more likely you were to sense trouble before it arrived. In the workshops, she found that, unlike the theoretical aspects of meteorology, engineering was all practice and was vital for anyone who had ambitions to fly their own plane. And she was thinking ahead by now; hiring a ship at the weekends would not do it for her, she wanted to be a real pilot, a professional, who was paid to fly, rather than a pupil who paid someone else.

On 15 March she went up with a new instructor, a young pilot who felt this girl needed taking down a peg or two. They were doing take-offs and landings and Bobbi, at the dual controls, took the ship smoothly up to about 100 feet. The instructor, sitting alongside, pulled the throttle back and she immediately put the nose down to counter the power surge. They were over the end of a good field and if she could make a three-quarter turn she'd have the whole length of it for a practice landing in rough conditions. However, Pop Burdett had always told her never to make a full three-quarter turn without having enough altitude, and she knew she didn't have that – she would have to do the turn in quarters. The instructor gave the plane more fuel and asked her why she wasn't using the extra push and executing the whole turn. She told him what Pop had said. He disagreed and took over the controls, bringing the ship round and back to the original position so he could show her how it should be done. Once more he accelerated and began to demonstrate the three-quarter turn. He was nervous, that was quite clear, and Bobbi was beginning to feel concerned: he really didn't have the height and she suspected he'd suddenly realised that. But he was a young man, and she wasn't, and he wouldn't back down, he simply couldn't. So he gave it more gas and kept the nose up and, as Bobbi knew backwards by now, if you gave a Curtiss Jenny gas, you kept the nose *down* because if you didn't, she'd go into a spin. And he kept the nose up and they went into a spin.

The first her Mom knew of it was reading the evening paper. The story read:

STUDENT SUFFERS CONCUSSION OF BRAIN WHILE PILOT ESCAPES UNHARMED. Miss Evelyn Trout, 22, from 412 South Soto Street, student aviatrix, suffered concussion of the brain and cuts and bruises when a plane at Burdett Airfield, where she was receiving instruction from Mr Dale Page, crashed a short distance to the ground after its motor stalled. Miss Trout was taken to Inglewood Emergency Hospital. Mr Page was unhurt.

Her mother got there just as Bobbi came round, and the first thing she said to her daughter was, 'Honey, you will give up flying, won't you?' Bobbi squinted at her through slit and swollen eyes and told her, 'No, I love to fly.' Her mom took her home and decided that if, as usual, there was no way to stop her daughter, she might as well try and make it all a bit safer. There was still family money in the bank and she decided to buy Bobbi a plane of her own.

A few weeks after getting out of hospital, Bobbi was back in the air and gained her licences; she was also the owner of her own plane, an International K-6 which her mother had bought from Burdett Airlines. Now she had the plane and she had the licence: what was she going to do with them? The first step was getting used to the K-6, a four-seater and both larger and more powerful than the Curtiss Jennies she had been trained on; the second was finding some answer to the material question: how does a woman with a plane make a living? First, she decided: be visible.

A local barley field was being converted into an airstrip for a series of air races to be held in autumn 1928 (the field was later to become Los Angeles International Airport). When the work was mostly completed and the strip usable, local and national flyers were invited to attend a gala opening. Charles and Anne Morrow Lindbergh were to be there, as was a young woman who had attained celebrity by being the first female passenger aboard a US–Europe Atlantic crossing. Though she might not yet have provided much actual flying to support her growing reputation, Amelia Earhart was, undoubtedly, the person to know in women's aviation and anyone with half a brain could see that she was going to become a major star, if not a major pilot. As it turned out, Bobbi didn't meet Earhart but she did get to shake hands with Colonel Lindbergh, who had a weak grip but, unlike Earhart, was a genuine legend by now and he could certainly fly, as he and his aerobatic team, the Three Musketeers, proved later in the day.

During the show, she was noticed by a local department store owner who hired her plane, the K-6, as part of an advertising campaign. She didn't get to fly for them, they just wanted the plane on the roof of the store, but she did earn money and get local attention. Sunset Oil offered her fuel and oil in return for painting their logo on the side of her aircraft. Her big chance came in December – a lucky time of year for her – when she landed at a local field and met R. O. Bone, a California aircraft builder. Chance or design, she didn't know, but Bone was looking for someone to demonstrate his new ship, the Golden Eagle, a high-winged monoplane, and a local aviatrix who was building a reputation seemed like the ideal saleswoman. He waited as she climbed out of the cockpit, introduced himself, shook her hand (there was nothing wrong with *his* grip) and asked her there and then, 'Would you like to demonstrate my new plane?' She asked what the pay would be. He told her, 'Thirty dollars a week and all expenses.' She said that sounded pretty good and she'd like the job. He asked when she could start. She said, 'Would today be soon enough?' And that was it; she was off to the factory for a course in construction and engineering, which was no problem considering her background at the garage and in woodwork class back in junior school, and on 14 December she attended the opening of Los Angeles Metropolitan Airport (fields were springing up all over) and took part in and won the air race held to celebrate the event.

She began to fly the Golden Eagle up and down the coast, dropping in at airports, letting local pilots get a good look at the craft, then offering them a flight. Sales weren't great – the Depression was easing itself over the financial horizon with a cheesy grin and money was becoming, if not short, at least harder to find. However, it did mean she had a good plane to fly and no longer needed the K-6, which she sold. Its new owner spun-in a few weeks later; he was unharmed but the plane was a write-off.

Bobbi had a living, she had a plane and now she found a plan. Local flyer Viola Gentry had set up an endurance record, spending eight hours aloft; she wanted to break it and thought the Eagle was the plane in which to make the attempt. R. O. Bone agreed with her and installed extra fuel tanks in the ship. Bobbi had no worries about her own ability to stay in the air alone; as always, she was a pretty good judge of her own capabilities: she knew what she could do and what she couldn't, and she also knew how far she could safely push herself without risking the Eagle.

She took off on 2 January 1929 and stayed up for 12 hours and 11 minutes. During the first hours, as she circled over the walnut groves, the apricot and peach trees, the plane was fuel-heavy and sluggish and demanded constant attention, but as the hours went by, the fuel was used and flying became easier; not that she was looking forward to a night landing (which would be her first). But when the time came, she put the Eagle down without a bump and found herself surrounded by a wall of light as the photographers' flashes went off in her face. There were cheers, there were questions, she was carried around the field on shoulders, there were autographs, there was a party later back at the family house organised by her brother Denny. There was also a note of encouragement from army flyer Carl Spatz (who later changed his name to Spaatz; one wonders why?) who was leading a forces team in a bid to gain the military endurance record. His superiors had told the army crew to get that girl out of the sky, but Spatz was a pilot and respected Bobbi's record. Besides, what were the army supposed to do: shoot her down?

The Women's Endurance Record put R. O. Bone, the Golden Eagle and its pilot right into the national headlines, and one of the readers was Elinor Smith, no stranger to courting publicity herself. Aged 18, she had flown under the bridges of New York's East River. She was also no stranger to male prejudice. After one record-breaking flight in the mid-1920s, her male co-pilot, unable to accept the situation, stepped up on to the podium and said that he'd flown the plane over the whole distance himself; Smith learned a valuable lesson from that situation and was more careful in future about who she flew with. After reading about Bobbi's record, she went up alone and beat it by an hour.

R. O. Bone had promised Bobbi that if her record was broken, he'd put up the money for another attempt, and he was as good as his word. In February, Bobbi took off again with a full load of fuel and began to circle Mines Field. Her intention was to fly through the night, the first time this would have been achieved by a woman; however, problems arose as it began to get dark: she simply couldn't stay awake. Her eyes shut, she nodded forward, the plane began to dive and only as the engine revs increased as a result would she wake up again. She began to sing, rub her face and neck, peel oranges (sponsored by Sunkist) and eat anything within reach – and this was what saved her from having to land. As she discovered after seeing her doctor, her body burnt energy at an unnaturally high rate and as a vegetarian she simply wasn't getting enough protein; he put her on a steak and eggs diet and her capacity to stay awake after dinner improved radically. Up in the air, as the minutes ticked slowly by, she found she'd run through every song she knew and everything edible in the cockpit but at last the sun came up, the needle on the fuel tank crept down towards zero and it was time to land. She'd been in the sky for 17 hours, 5 minutes and 37 seconds. In the crowd to welcome her this time was Will Rogers, national newspaper columnist, film star, international wiseacre, a half-Indian entertainer and writer who had managed to capture the nation's attention with his down-home common-sense style of wisdom. He also happened to be wild about flying and was a good man to have on your side, particularly at a time when the whole country was going plane-crazy. He wanted to talk; all she wanted to do was sleep.

Another new airport was opening – Grand Central – and among other events, there was going to be a women's pylon race. Bobbi thought she could give it a go and get a chance to meet a few of the film stars and flyers who'd be attending the event. In the race she placed third but got to meet and become good friends with the two other competitors, Margaret Perry and Pancho Barnes. Altitude record holder Louise Thaden was also around in her Travel Air, and she and Bobbi discussed possible collaboration on a long-distance flight. Marvel Crossen, who'd been earning a living flying in Alaska, was visiting, although not taking part in the competition, and Bobbi got to spend some time with her, discussing the problems of running an air taxi service in extreme conditions.

Bobbi's financial needs were being well taken care of through sponsorship – Richfield Oil was not only donating their products free but paid $1000 for the privilege of painting their logo on the Eagle. Taking advantage of the public mood and the arrival of a new Golden Eagle, the Chief, which Bone had been developing, Bobbi decided to use the ship's extra horsepower to try for the women's altitude record. In June 1929 she reached 15 200 feet, claiming the light aircraft record. But her thoughts were elsewhere. For months now there had been rumours that various interests in California had been trying to set up a women's cross-continental Air Derby, and now the rumours were confirmed. The race would start from Clover Field, Santa Monica and would finish in Cleveland, to coincide with the opening of the national air races. There was prize money of $25 000 to be won in two classes of planes, light and heavy.

Bone was in agreement with his chief pilot: no woman flyer could possibly refuse the challenge of the Derby and if she placed well, the Golden Eagle would receive a terrific boost. If she didn't, just taking part would underline the ship's good qualities. They did decide, however, to fit a new engine behind the cowling, a Kinner 100 hp which could achieve a speed of 120 mph. More speed made good sense, Bobbi thought, but she would have liked a bit longer than a day to get used to the new motor; there were also a whole raft of would-be sponsors, from whom Bobbi chose the National Fuel Company for her gas supplies. While she was waiting for the engine to be fitted, she renewed acquaintance with flyers from the Grand Central airport opening, most of whom had entered. Pancho was busy shocking everyone, checking out her plane with a cigar between her teeth, causing one gentleman to announce, loudly, that if that was what women flyers looked like, then he'd never allow his daughters to leave the ground. Pancho's response is unrecorded. Bobbi also the got the chance to meet Amelia Earhart, whose star was still climbing, under the capable guidance of her publicist, G. B. Putnam. The two women got on well and the talk turned to the possibility of shared endurance records. Bobbi had been disappointed when earlier plans to fly with Elinor Smith had been kyboshed by R. O. Bone on the grounds that it-didn't-fit-into-his-sales-scheme. Earhart was a different matter, her name was gold in the bank, and if Bobbi wanted to endure with Amelia that was fine with him. They agreed on a possible date, the only problem being Earhart's crowded schedule, which she promised to check with Putnam.

The Derby started at 2.00 p.m. on 18 August. It was a landmark for women flyers but it was also a trade race: just about every ship that took off from Clover Field that day was carrying the sales hopes of some or other company and not a few, it was hinted, were also carrying the results of sabotage (whether this was inspired by business competition or the anti-women-in-the-cockpit lobby was unknown). Not that Bobbi cared a lot either way, unlike Earhart, who always seemed to be on a mission for women in the air. Her job was to fly the thing and, if possible, win the light aircraft section of the race. Her intense practicality never had much time for causes or crusades; there were situations and you made the best you could out of them; if there was nothing to be done, then you accepted that and didn't waste time grieving over it, you just got on and found something you *could* do. In a lot of ways it was an attitude she'd inherited from her wandering, ever-inventive father; luckily, she'd also got a big dose of her mom's practicality. The two together were a formidable combination; but not quite formidable enough for the official photograph of the Derby contestants. Whether it was the photographer, R. O. Bone or Bobbi herself, for the first time in years she appeared in public in a skirt, along with every other woman in the race – except Pancho Barnes.

She took off fifth and from the moment her wheels left the ground, she could feel the power of the new engine and knew she had a real chance. Call it hubris or bad luck or anything you like, but hardly had the thought surfaced than the oil pressure gauge fell. She swung away from the official race course, avoiding built-up areas, trying to ensure that, should she lose power, there would be open space nearby for an emergency landing. She was nursing the plane along when, quite suddenly, the problem vanished. It must've been a bubble in the line, she reckoned, and pulled back on course for the first leg finish at San Bernardino.

There were questions being asked by the competitors about the next day's route; particularly the midday stop, which many considered unsuitable. The race organisers were adamant: there would be no change of the race course. Bobbi went to bed early that night, not realising that Pancho Barnes was about to join the fray. When she woke in the morning, late – the hotel forgot her early call – she discovered that Pancho had prevailed, the route *had* been changed and the other racers were already at the field, preparing to take off. She grabbed a cab and raced out to the airport, getting into the air well behind the field. They were heading for Yuma, the new midday stop, and she was pushing her engine hard, getting every rev she could out of it, which might have been why she ran out of fuel five miles short of Yuma. Or maybe somebody had been tampering with the tank; either way, the stick was dead in her hands and she was going have to come down fast. She spotted a ploughed field, which was just over the border in Mexico, but she didn't know that; she didn't know, either, that the furrows

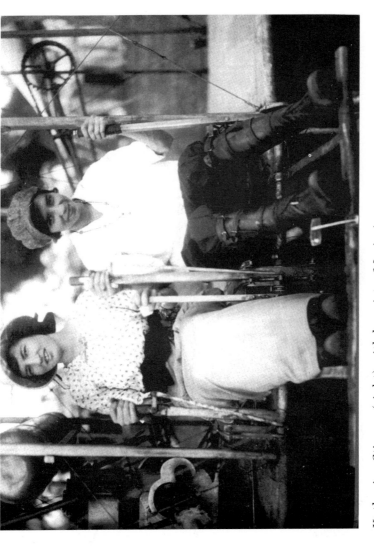

Katherine Stinson (right) with her sister Marjorie
Reproduced with permission of the Center for Southwest Research, General Library, University of
New Mexico. Neg.000-506-0125. Photograph by Anderson, KC.

Lady Mary Heath

Pancho Barnes
Reproduced with permission of Corbis Photo Library © *Bettmann/CORBIS*

Bobbi Trout
Reproduced with permission of Corbis Photo Library © *Bettmann/CORBIS*

Bobbi Trout
Reproduced with permission of Cheryl Baker. Photograph by Stockton of Hollywood.

Bessie Coleman
Reproduced with permission of Corbis Photo Library
© *Underwood & Underwood/CORBIS*

Hanna Reitsch
Reproduced with permission of Corbis Photo Library
© Hulton-Deutsch Collection/CORBIS

Amelia Earhart
Reproduced with permission of *The Flight Collection*

Amy Johnson
Reproduced with permission of *The Flight Collection*

Chubbie Miller
Reproduced with permission of Corbis Photo Library
© *Underwood & Underwood/CORBIS*

Bill Lancaster and Chubbie Miller
Reproduced with permission of Corbis Photo Library
© *Hulton-Deutsch Collection/CORBIS*

were extra deep and uneven, and when her wheels touched, the plane flipped and she ended up on her back. Undoing her harness, she slipped out and checked the ship; it was going to need a lot of work.

It was more than three days before the plane was trucked to Yuma where R.O. Bone and his mechanics could get to work on it and put the damage right. Bobbi took off well behind the field but she had no thoughts of pulling out of the Derby while her ship could fly – she gave it extra fuel and arrived in Kansas City to find that most of the other pilots were long gone. She also heard the news about Alaskan pilot Marvel Crossen, who had been killed during the Yuma–Kansas City leg, parachuting from her plane when the desert heat and the poisonous exhaust fumes from her engine became unbearable.

Although out of sight, the competition wasn't *that* far ahead and she thought she still had a chance of placing. She left Kansas City at speed and pushed the plane to its limit and beyond. An ignition switch failed, and she found herself once more guiding a powerless craft down on to the nearest convenient field. This one, at least, was flat but small and she had to turn sharply, catching the tip of her wing on a fencepost, tearing the fabric. She'd never been a needlewoman like her mother but, when she had to, she could darn a pair of trousers or fix fabric, and she did so now, using a tin sheet as the patch and baler twine as thread. She fixed the ignition and took off for Columbus, where the race officials almost disqualified her for the unconventional state of her plane. She didn't see anything wrong with it – it seemed to be handling better than ever with the tin in place.

Leaving Columbus virtually in sight of the tail-end of the field, she got to Cleveland having overtaken two other contestants, and felt pretty pleased with herself. Later in the week, after the air races, a banquet was held for the Derby pilots; Bobbi had lost her luggage during one of her enforced stops and went in trousers and a white shirt borrowed from Elinor Smith; everyone else was in evening dress but she managed not to feel too uncomfortable. When she returned the shirt, she and Smith resumed their discussions on endurance flying; they both felt that with two pilots and mid-air refuelling, the time aloft could be pushed to far greater lengths than anyone had conceived of so far. Besides, working together would save them from the leap-frog competition they'd become involved in.

On the last day of the races a group of pilots was standing around in the shade under the grandstand, chatting about this and that, the events of the race, the possibility of sabotage, what they might be doing over the next few months, when someone said, 'Wouldn't it be nice if we could all get together like this and talk more often?' And somebody else said, 'Yeah, why don't we have a women's organisation?' Then Bobbi said, 'That means a lot of red tape and by-laws and so forth,' and Earhart said, 'Bobbi, why don't you let us work on that back east?' And Bobbi said, 'Why not?' Which is one story about how the Ninety-Nines association of Women Pilots was born; right there and then under the bleachers.

Flying back to California, Bobbi gave a lift to Jack Helms, an old dancing partner and distributor for the Golden Eagle Company. Outside St Louis, while Jack, who was also a pilot, was flying the ship, they saw a huge black cloud boiling up in front of them; it clearly indicated rough weather and since neither flyer was in a hurry, they decided to land and wait out the storm. But the weather was impatient; rain began to lash at the ship, getting heavier until they could barely see out of the cockpit screen. Jack took her down and landed in the nearest field, which the rain had turned to liquid mud. The wheels caught, the plane upended and smashed nose-first into the earth. The pilots were unharmed, but the Golden Eagle was a write-off. Bobbi and Jack continued the journey by train but almost as they travelled, the national financial outlook got darker and the Depression edged that much further into the nation's life. R. O. Bone was no fool and could see as clearly as anyone that building small planes for private owners was not the business to be in right now. The company closed down, Jack Helms was out of a job and Bobbi was out of a plane.

Then Elinor Smith called. She was really enthusiastic about the idea of a long-haul record attempt. Bobbi invited her to California; there was a chance they could make that attempt as a local promoter, Jack Sherrill, was ready to put up a plane and support services. Ordinary people, depressed by the economic slump, were more than ready for something adventurous and entertaining to take their minds off their troubles, and Sherrill had organised a radio link with a local station so those on the ground could follow the adventures (or misadventures) in the air. For the flight, he was offering a Sunbeam biplane fitted with a 300 hp Wright Whirlwind engine, with a Curtiss Carrier Pigeon acting as mother-ship for fuel and supplies.

Smith travelled across the continent, taking planes by day, trains by night and met Bobbi at Grand Central Airport on October. The two went to see Sherrill and told him they would need at least one practice run to check out the various procedures, particularly the refuelling; he agreed, and set it up for the next day. Back home at the Trouts', while her mother cooked supper, Bobbi and Elinor tried on the suede flying suits that had been specially created by the finest tailor on Wilshire Boulevard in LA for the attempt; Bobbi's was fuchsia, Elinor's green. They also tried to decide who would pilot the ship and who take care of the mid-air transfers on the trial flight (they would swap four-hour stints on the actual attempt). They couldn't and, in the end, they flipped a coin: Smith was to fly, Bobbi to refuel and provision.

Next day at the airfield, Bobbi went over the arrangements for receiving the fuel; a rope would be lowered with provisions in a bag weighing it down. She would catch the bag and reel in the rope, which was tied to the nozzle of a fuel pipe. This would have to be secured to a port which opened into the cabin fuel tanks. Once filled, fuel would be hand-pumped from these tanks into the smaller engine tank. There would be two refuelling links a day, each transferring

180 gallons. It all looked simple enough on the ground. At a few thousand feet above the earth, Bobbi reckoned it might be another story altogether. And that's how it turned out. As the two planes flew in circles above the airfield, Bobbi was able to grab the bag of provisions and pull in the fuel line; the refuelling itself was going well when the planes started to drift apart. The line jerked out of the cabin tanks and doused Bobbi with petrol. As she gasped, she swallowed and choked. Smith took the Sunbeam down fast and Bobbi was hurried to hospital and placed in an oxygen tent. She wasn't there long; her main concern was getting back to the ship and making sure that the same thing didn't happen again.

It didn't. They took off and performed the first fuel transfer perfectly; however, the Sunbeam was flying tail-heavy; whatever Smith tried to counteract it made no difference: the craft was inherently unstable. They conferred, as much as the engine noise would allow, and decided they'd have to land and counteract the imbalance. Once down, they checked out the loading and came to the conclusion that it had to be the radio equipment – two huge and heavy boxes. Much as they hated to lose their ground contact and their radio audience, it had to go. They made a second start, and a second refuelling went wrong when the rope got caught and Bobbi had to free it by hand, suffering rope burns. They came down once more and she sensed that Smith was ready to call it a day. She agreed; a night's sleep would set them up for the morning. They took off for the third time on 27 November, this time determined to see it through.

They lifted off in the early morning, the Sunbeam flying perfectly, and set up their circular course around the airfield. Smith was flying the first shift while Bobbi checked out her tasks: the engine oil had to be changed every day, the rocker arms on the engine needed to be greased and fuel pumped from the cabin tank to the engine tank, a task that caused pain in the arm muscles after the first few minutes but had to be performed hour after hour, day after day. However, the refuelling went smoothly for the first two days and Bobbi worked out a system of coiling the rope that would avoid both tangles and rope burns, just as long as she kept a weather eye out on her own position. On one refuelling she noticed that the spare rope had wound itself around her leg: she would have been carried away had she not unwound it frantically as the mother-ship began to pull clear.

On day three, as the supply plane lowered the fuel nozzle, Bobbi noticed black smoke streaming from the exhaust. It looked like they had a fire aboard, just as gallons of highly inflammable aviation fuel began to flow down the pipe. Smith had seen the smoke too and as Bobbi ripped the hose out of the cabin tank and let it swing free, Smith pulled rapidly out of the shadow of the mother-ship as it began to lose power and sink toward them. Flying clear, they watched it go down, trailing black smoke, as the pilots frantically fought to pull it out of the dive. They finally managed, sweeping low over the ground but the hose was still trailing and caught in a fence. Seriously unbalanced, now trailing a few hundred yards of fencing, the Curtiss lurched on for a couple of hundred yards and then landed. The pilots clambered out and waved up at the Sunbeam. They were unhurt, but it was clear that their plane would not be taking off for a while. Elinor and Bobbi decided to fly on for as long as fuel reserves allowed. They landed at last after 42 hours, with fuel gauge on zero.

The crowd was ecstatic: this was the world's first refuelling endurance flight made by women, they had all the publicity that they or Sherrill could possibly want. Bobbi was proclaimed the Sweetheart of Los Angeles. Sponsorship offers came flooding in and she was presented with a specially designed leather flying suit heated by electricity. She got fur-lined flying boots as well but it was the suit which evoked most interest. News of its properties eventually reached Colonel Charles Lindbergh, who was planning new flights with his wife Anne Morrow. Since both of them suffered from the cold, he wrote to Bobbi asking if she would share the pattern and design of her suit. She was happy to do so and the Lindberghs had suits of their own made, using canvas, for lightness, rather than leather.

Elinor didn't stay around for long after the flight. Bobbi detected a certain coldness in the other flyer's attitude. There was nothing obvious but maybe, she thought, just possibly Elinor was annoyed that Bobbi's name had been painted on the fuselage of the plane before hers and that, as the local girl, she was getting most of the press coverage. Not, after all, so small a matter when sponsorship was becoming ever harder to find.

As always, raising money was a problem but not yet an insuperable one. Bobbi began racing, winning prizes at local meets, including the opening of Burbank airfield. Public fascination with aviation and aviators was still high, and Amelia Earhart was becoming a national hero, proving that women flyers were of particular interest. This meant, in Bobbi's mind, they were more likely to gather in what sponsorship there was now that the Depression had really taken hold. And the one sector, given that she was unlikely to find backing from the bootleg liquor industry, that still had cash to spare was the film industry.

Many flyers worked as stunt pilots – Elinor, Ruth Elder, Pancho and Mary Wiggens, who Bobbi had taught to fly, had done so – but Bobbi reckoned the risks of that job were a little too high. Fashions were changing and films about flying were becoming less popular; directors were demanding increasingly crazy and dangerous stunts from their pilots. All in all, publicity seemed a more likely route; she had already utilised the celebrity she'd gained from the flight with Smith, writing to local companies offering her services and her plane, though since selling the K-6 she'd been forced to hire a ship whenever she needed one; another expense. She also looked at business ventures of her own, setting up a company to fly seafood from the East Coast to California, though this fell through at the last moment when the financial backer died of a heart attack. A game and poultry air delivery service was also mooted, which stumbled at the last hurdle, but Bobbi had never been one to get depressed if something didn't work. She was intensely single-minded and driven as far as her own ambitions were concerned, but was equally practical. She was not a woman who lived her dreams in her head; she had always needed to turn thought to action, whether it was in building a flat-bottomed boat at the age of six or persuading her parents to buy a service station aged 19. She ensured that her common sense stopped any flights of fancy before they left the ground, while underpinning her career as a pilot in the real world. When that was no longer possible, she would turn her talents to something else without more than a moment of regret.

In late 1930 she got a call from Edna Mae Cooper, an actress at MGM Studios who had appeared in Cecil B. DeMille's *Male and Female* in 1919, and many other films but had not so far managed to achieve star billing. She had, however, just earned her pilot's licence and as her movie career wasn't going brilliantly, was looking round for some way to utilise her flying experience. It was practically nil, but that had never stopped an actor! Cooper had met a Hollywood investment broker, Joseph Martin, who was interested in advertising the new LA office he was opening and thought an endurance flight with a pretty girl at the controls might be just the thing. He was also a practical man and realised that an endurance flight with Edna Mae Cooper at the controls would probably endure no more than five minutes; he told her that if she could find an experienced woman pilot to fly with her, he'd sponsor the attempt.

Bobbi agreed. Her aim was always to get back in the air if it was at all possible. She met with Edna Mae and Joe Martin and talked through the possibilities. Martin was to hire a plane and install the necessary fuel tanks, have the engine thoroughly overhauled and make all the arrangements for refuelling. Bobbi would fly for as much of the time as possible and Edna Mae would take over for rest periods and look after maintenance and mid-air transfers. For the record-attempt ship Martin went to Corliss Mosely at Grand Central Airport who provided a Curtiss Robin monoplane. Mechanics were hired in to check the engine and install cabin tanks and hand pumps for getting the fuel to the engine. He called the plane *The Lady Rolph*, after the wife of State Governor James 'Sunny Jim' Rolfe Jr. who was writing his name in the history books by putting a tax on food and endorsing a brutal prison lynching. However, in terms of courting publicity, it was a shrewd move, bringing the attempt to the attention of the press. The mother-ship would be the same Curtiss Carrier Pigeon which had accompanied Bobbi's previous attempt (with all the faults fixed, she hoped).

Martin and Edna Mae used their Hollywood connections to invite stars and starlets to the take-off on 1 January 1931. There were flowers and cameras and crowds and good wishes a-plenty; Bobbi was more concerned about good mechanics; the engine had been checked out, new rocker arms added for ease of oiling and a wind-driven fuel pump fitted, which was gong to make things a lot easier, but as she knew, everything in the air always seems OK until the moment it isn't.

They took off with half-empty tanks, a far easier manoeuvre than dragging a fully loaded craft into the sky. The mother-ship would be following them up so they'd be able to transfer fuel in the air but, as they circled over the field, looking down at the crowds, they could see a little group gathered around the Curtiss Carrier Pigeon's prop, trying unsuccessfully to swing it and get it started. It began to get dark and Bobbi knew she had to make a decision: landing later, with virtually empty tanks, would be tricky and might result in damage that would stop an immediate restart. If the Curtiss did manage to get off the ground, a night-time fuel transfer would be equally tricky, particularly with only one experienced pilot. She decided to go down at once; it was the sensible thing to do. Joseph Martin didn't agree; he was furious, swearing with a freedom Bobbi did not and had never liked. A man or woman should be as in control of their own tongues as they were of their lives, she reckoned. Martin bellowed that he already arranged for a smaller plane to come up to them and lower five gallon cans of fuel on a rope. Bobbi suppressed a grin of sheer disbelief and asked just how they would see these cans in the dark. Martin said there'd be a lighted lantern tied to each one; at least, there would as soon as he could find a pilot willing to do the job. The scheme was too stupid to get angry over. Bobbi told him to find another mother-ship and fit it out properly and only then would the record attempt begin again. She would have been wise to extend her caution over Martin's slapdash preparations to her own plane but there was a lot to do and, after Martin's promise that everything would be ready for 4 January, she let it go.

'GIRL FLYERS TRY AGAIN FOR NEW ENDURANCE RECORD. Up there again!' the newspaper trumpeted as *The Lady Rolph* headed down the runway for the second time, this time fully loaded and the very devil to pull over the high tension wires beyond the airfield boundary. They got up and began to circle the field. After a few hours the mother-ship took off and the first refuelling went without a hitch. Oil cans were lowered as well as mail and provisions. A note asked the aviatrices to tie the cans to the small parachute that was attached and send them back down for reuse. Bobbi wasn't so sure; this looked like another example of Joseph Martin's inexperience, but Edna May was handling the logistics and it was really her call. Bobbi just shouted through the engine noise and told her to make sure she threw the chute well clear of the plane. Edna Mae nodded, not really getting it, and hefted the bundle as hard as she could. It wasn't hard enough, and she understood exactly what Bobbi was worried about as the chute unwrapped in the slipstream, billowed out and got caught around the tail of the ship, exerting a huge drag which unbalanced the whole craft. They had full fuel tanks which added to the overall weight and, if they went down, these would almost certainly ignite.

Bobbi wasn't dying for anyone. She wrestled with the controls, trying to keep the plane on an even keel against the natural tendency of the drogue to make it flip. They began to lose height. Edna Mae crawled down inside the fuselage and tried to reach the ropes of the chute and pull them free. It was clear this wasn't going to work. They were now flying only feet above the runway and while the ground crew could see the problem and would, hopefully, come up with a solution, those high-tension wires that had been a concern on take-off were now looking even more dangerous as the fully-loaded plane hurtled towards them. Bobbi wasn't thinking, she was simply flying, drawing on every skill she'd ever learnt and a few she hadn't. The ship lifted and cleared the wires by inches; the chute, whipping wildly in the slipstream, was beginning to wind itself into a plait, still a danger but the drag was reduced, and they were able to carry on gaining height.

The mother-ship followed them up and a long bamboo pole was lowered with a sharp knife attached to the end. Edna Mae took it, crawled down the fuselage and eased herself out of a tiny access trap, leaning into the wind to start cutting. Both women knew that should she miss and cut into the fabric of the aircraft, they could lose control and go straight down. As the moon came out and the night got colder, Edna Mae sawed her way through cord after cord. It seemed to take hours of work but eventually the last was cut – there was no chance of a shouted warning – and the plane leapt forward. Bobbi throttled back, Edna Mae crawled out of the wind; they were safe for now. Or so it seemed.

After a few hours, it was Edna Mae's turn to take the controls while Bobbi rested. The two women exchanged places and Bobbi lay down to sleep on top of the cabin gas tank. Hardly had she shut her eyes than she opened them as she slid off the tank. The plane was diving. She yelled at Edna Mae to haul back on the stick. The novice pilot did as she was told and hauled back. The plane righted itself, then began to climb so steeply that Bobbi feared a stall. Push forward on the stick, she ordered. And down they went again. By now she was fully awake and in a position to take over the controls and get the plane level. There appeared to be no reason for Edna Mae's eccentric flying but on questioning her it suddenly became clear: in the dark she had no feel for the horizon, something most pilots seem to have by instinct. It was going to be a long flight indeed, Bobbi reckoned, if Edna Mae could only take over in the daylight.

Their third day aloft, 7 January, was Bobbi's 25th birthday, and a chocolate cake was sent down from the mother-ship with the supplies. After sharing a slice each (there were no candles) the crew settled back into tasks that were becoming almost monotonous by now. They were flying at a steady 50 mph, running at 1350 rpm, keeping to a height of 2000 feet; they needed to keep an eye on the fuel tanks (one in each of the wings) and keep them filled from the big tank in the cabin, as well as check the oil pressure and keep the temperature steady. Morning and evening, one of them would change the engine oil, opening a tap to let out the old oil, tipping in the new to keep the whole thing lubricated; then there was the greasing to see to and galley duties; not that these extended to much beyond handing up a sandwich or orange juice. There was no radio, as in the first flight, as the weight would have been too much; they relied on notes dropped down with the supplies for any information and on the night of the third day they were informed that a storm was blowing up and advised to fly down the coast to Imperial Valley Airfield and circle there until local conditions were clearer.

By 8 January they were back over Mines Field, flying in an easy circle as the afternoon turned into evening. Edna Mae was off duty, sleeping in the cabin and Bobbi was in the pilot's seat, relaxed, keeping one eye on the fuel, temperature and oil gauges; it seemed, for a while at least, that everything was working the way it should. She looked through the screen at the countryside below. As it got darker, the shadows creeping across from the mountains, she saw car headlights come on and house lights go up. There were not all that many; electricity hadn't yet conquered the world, so she could make out individual cars and lorries, farm houses, service stations, the glow of cities in the distance, the lights on Mines Field . . . And then, as it started to get late, the lights, one by one, began to go out.

Houses and streets, the airfield, even the headlights of cars following the winding roads, all vanished until it seemed as if there wasn't a light anywhere and the plane was utterly alone in the dark, enfolded by a huge silence. Even the engine noise had faded, and there was only the vast splash of the Milky Way above and . . . nothing, absolutely nothing. It was a moment she would remember forever, she knew that, as near to heaven as she'd get without buying a ticket and making the trip. She was happy to sit, just nudging the stick, as she and the earth turned round through space and the sun began to glow behind the mountains and the ninth day dawned. And out of the clear blue sky, black oil began to squirt over the cockpit screen.

She checked the rev counter: the needle was flickering wildly; the cockpit screen was getting covered with the thick, oily sludge; more oil was escaping from the engine within the cockpit, covering her. Edna Mae came crawling up from the cabin but there was nothing she could do. Bobbi was far better qualified to fly in these conditions. They discussed whether they should go down at once or see how the situation panned out. Bobbi reckoned she could keep the Robin in the air for a while yet, so they decided to keep the first refuelling appointment of the day. Edna Mae went back down to get ready for the transfer. When it came, the mother-ship had to take care of the manoeuvres, since it was as much as Bobbi could do to keep their craft level. Refuelling took place but oil continued to stream from the engine. Bobbi knew they would have to land soon but she wanted to stay up as long as was safely possible; they'd put in so much time already that it seemed a shame not to give it their best shot. Besides, they now had full tanks.

She reckoned the problem might well stem from a crack in one of the engine cylinders. The nature of the leak, with the revs falling as the oil spurted, then when the leak stopped, increasing until they reached a point where more oil appeared and they fell again, suggested something of this nature. It would be a recurring cycle which Bobbi felt she could cope with for at least a few hours, but not after dark. She flew on, conscious now, as she had not been during the night, that she was sitting on the buckle of her parachute harness and that it was extremely uncomfortable.

By mid-afternoon, nature was beginning to make itself felt and Bobbi realised she was going to have to use the can (literally, a 1 lb coffee can under a wooden seat with a hole cut in it, back down the fuselage). The internal oil leak was more or less under control, though the external leakage was still there and the revs were fluctuating wildly. She called Edna Mae and told her to keep the ship level, and made her way back down to the facilities. As she pulled down her trousers the plane began to dive. She shouted through, 'Pull on the stick, Edna Mae!' The ship settled back and she settled down. Then up went the nose. 'Stick forward, Edna Mae!' Then down went the nose. 'Stick back, Edna Mae!' By this time the absurdity of her situation had struck home and she gave way to helpless laughter, hanging on to the seat as they switch-backed across the sky. The oil covering the cockpit screen had left Edna Mae in the same situation as night flying; she simply couldn't find the horizon and keep the ship level. Bobbi finished as quickly as circumstances allowed and took over again. She told Edna Mae that they'd be coming in to land soon and if she wanted to make a good entrance, she'd best get ready. The actress needed no prompting. When they landed, after 172 hours, and 50 minutes flying, Edna Mae stepped from the plane like the star she'd always wanted to be; Bobbi stepped out like the pilot she was.

When the engine was stripped down, the problem was found to be a crack in no. 1 cylinder. Without that, they could have stayed up for a month. Union Oil, who supplied the attempt, as well as covering Bobbi, presented the flyers with a trophy; the two were popular personalities at events all over the nation and Bobbi soon needed to get a supply of business cards printed. King Carol of Rumania, who was an air enthusiast, presented her with the Aviation Cross, an honour she shared with Lindbergh and Earhart, and that lasted somewhat longer than the Rumanian monarch, who gave up his throne for the love of a red-headed dancer. Other awards followed, both from America and Europe.

Bobbi was getting interested in setting up a new long-distance flight, this time one that went somewhere: from Hawaii to California. For a while she worked for the Cycloplane Company of Los Angeles, where she taught pupils to fly the Cycloplane, a craft designed to enable any pupil to master flight by easy stages. Her favourite pupil was a young woman, Mary Wiggins, who had earned considerable fame by jumping from a dizzying height into a tiny tank of water covered with a layer of burning oil. Her jumping career ended when she hit the bottom of a tank with her head and damaged her back. She was now stunting in films and looking to broaden her skills.

Other money-earning opportunities arose through advertising for, among others, Walt Disney, and there are photos of Bobbi in her plane, infested by mice with large ears. There was also a chance to fly to Italy where she was to make a parachute jump over Florence in honour of the memory of the nurse Florence Nightingale. The Pennzoil Oil Company agreed to underwrite the venture and Bobbi was ready to start when she stepped down in favour of a friend, fortunately as far as she was concerned; the plane flying the party across to Italy vanished with all on board. However, her main interest remained the trans-Pacific flight. It was planned to set off from Honolulu in a Lockheed Altair, specially adapted for long-distance flying with tanks for 450 gallons of fuel and a cruising speed of 180 mph. The craft was being built for the flight at the Lockheed factory and the attempt was being backed by Gordon S. Davison, a soap manufacturer. The team were offering advertising spaces on the plane from $100 up to $2500 with the promise that after the flight the adverts would be cut from the fabric for the advertisers to display as they wished. The arrival date was planned to coincide with the Los Angeles Olympics, but the flight never happened; Bobbi and her backers simply couldn't raise the necessary finance. It wasn't until Amelia Earhart, with her highly professional publicity organisation, turned her interest to a trans-Pacific flight that the distance was achieved.

Bobbi then found herself joining up with the Women's Air Reserve, started by Pancho Barnes. The last Bobbi had seen of her old friend had been a few months before at a local airport when Pancho, impatient to be on her way, and unable to find any ground crew to spin her prop, had turned on her engine and spun the prop herself, neglecting to put chocks in front of her wheels. Once the engine caught, the plane began to move off down the runway with the furious pilot running along behind trying to catch up. Typically, Pancho had styled herself General Barnes. Bobbi was to be Captain Trout. Mary Wiggins, the flame diver, became the Reserve's drill instructor and Bobbi was to spend many pointless, she thought, hours marching up and down a parade ground. However, the WAR did appear at the Cleveland Air Races to considerable effect in their smart powder-blue uniforms, black boots and berets.

Nobody was quite sure whether it was the WAR, the uniforms or just an eye for publicity that led LA Police Chief James E. Davis to create 'a special police squadron of commercial and highly trained amateur women pilots who held police appointment and could be summoned to duty in any situation requiring expert flyers.' The squadron was never called into action but Bobbi was able to use her connections to gain entry to the police shooting range where it became clear that, had she ever needed to pull a pistol, she could have done so with speed, accuracy and, since no champion rooster had ever accused her of being a supporter of the anti-gun lobby, deadly effect.

In many ways, an age was coming to an end. As the Depression deepened and Prohibition was repealed, Americans found themselves looking at the bar to drown their sorrows rather than looking at the sky to raise their spirits. Bobbi bought herself a new plane, a Steerman, but jobs were drying up. There was still a little advertising work to be had, some air-taxi jobs, but by and large amateur flyers were having to go greater distances and take far greater risks to get noticed. Flying was becoming more serious as a war in Europe looked ever more likely and it was from the aircraft factories of California that Bobbi's next idea sprang.

She noted that aircraft manufacturers were throwing away thousands of surplus or below-par rivets every day. It was a problem her father would have loved and, like him, once she put her mind to ways of reclaiming and reusing the rivets, Bobbi soon found an answer. Unlike her father, she also raised the finance and got plant built to do the job and as US factories started gearing up to war production, the Aero Reclaiming Company's product became of national importance. Without Bobbi, Rosie would have been only half the riveter she was.

Flying drifted away – not without regret. When she sold her last plane, the Steerman, she felt a real pang, but there were always so many other things to do. In the 1950s she invented a plastic eggbox, and opened a woodwork business; she became a skilled photographer, and sold property; she bought land and farmed it; she panned for gold; she bought businesses and ran them; she went back to architecture and designed a house. And, as time went by and whole new generations of young women began to fly, she found herself being continually accosted and told, 'If it wasn't for you, we wouldn't be here.'

She wasn't so sure about that. She had been *there* but, unlike Earhart and some others, she'd never carried a torch for a cause; for friends, yes, there were many torches but she had always been a strictly practical woman and mistrusted theories that didn't immediately pan out. And it was true – she was a survivor who refused to be daunted by prejudice or fate, and she flew on into history, dying in January 2003 at the age of 97.

6
'They say I shot a man and killed him'

*R*eally it should never have happened; the whole thing was an accident. Not the death, of course, that was quite obviously murder or suicide, but the rest of it, from the beginning, it all depended on a couple of words spoken a few years before at a party in London to a man she'd never met.

Jessie Beveridge was born in Western Australia. Her great-grandfather was an English bishop, her grandfather a clergyman, her father a bank manager. Her mother was a mother; in those days provincial Australian society did not look kindly on unconventional behaviour in the families of those who managed its money. Besides, she was also a clergyman's daughter and not given to expressing her wildness; that would be passed on to *her* daughter.

Jessie's upbringing was strict but loving. Her father was a man of great charm and the bank always sent him to open new branches, knowing he'd be able to make friends with and, more important, gain the trust of potential customers. Jessie herself wanted to be a concert pianist when she grew up and she practised hard, when she wasn't playing with her brother Tommy, who was also her best friend. He wanted to join the Navy and as soon as he was eligible he went away to the Jarvis Naval Academy. By this time the family had moved to New Zealand where the bank wanted to put down new roots. Jessie went to the Timaroo Girls School where she did well, particularly at her music. It looked like there might be a real future for her there.

Jessie and Tommy often talked about the things they'd do when they were grown up; the travels, the adventures they'd have together, the people they'd meet . . . and now Tommy was sending home photos of himself in his Navy uniform and it looked like he would be setting off soon on his voyages. Only, it didn't happen; he got cerebral meningitis and he died. He was only 21.

It must have done something to Jessie, made her think about the world that stretched beyond the long sermons in church, and she thought to hell with it and kicked her hat over the mill. No more church for her – the family was back in Australia now, in Melbourne, and that was a great city for a young woman who wanted to see if she had any wings to spread.

She gained a nickname, Chubbie, although she wasn't, she was small, hardly five foot, and slim and pretty and had all her father's charm. Then she met a man called Miller who wrote for the newspapers and she married him there and then, which her family thought was a mistake and they were right, in a way, but wrong in a way too. She was 18; she didn't know what she wanted, but she did know exactly what she didn't want, which was everything she had.

Miller was nice enough. They got on, and his was a different world where the things that appeared in the newspaper weren't just stories, they were part of everyday life, and she liked that feeling of being there, in the immediate present. There was no great romance but Chubbie was always a practical woman with determination to do better.

After a year or so, she and Miller drifted apart. For all his newsroom gossip, he was as young and naive as she was; there was no mention of divorce; back in the early 1920s that wasn't a subject to be taken lightly, but there was always amicable separation. Mr Beveridge, who knew a thing or two about human nature (only a fool thinks he can lie to his doctor or his banker) thought it might be nice if his daughter had a look at the Old Country: took a tour through Great Britain, visited Kent, where the family came from and there was an aunt with two sons to meet, spent a while exploring London. It seemed like a good idea, and if Miller couldn't get away from his job, then that saved a lot of discussions and saved a little pride for the man whose new bride was leaving so soon. She would go for six months, she would buy her own ticket, he would give her an allowance.

She found a flat in Baker Street sharing with another Australian girl, a friend of hers who'd come over at the same time. They saw the sights, got to know a few people, met an Australian dentist (there is, after all, no part of the known world that doesn't boast an Australian dentist) who invited them to a swanky party where there would be artists and entertainers from the stage. Naturally they went and it really *was* a party, and Layton and Johnson, who were about as big an international act as you could get, were sitting at a piano playing and singing their hits, '*ain't she sweet, a-walking down the street, I ask you very confidentially, ain't she sweet,*' and the dentist pointed out a tall fellow across the room and said, 'That's Flying Officer Bill Lancaster, he wants to fly to Australia. You should meet him.' So Jessie did.

He'd recently left the Air Force and they weren't that sorry to let him go. He was a good pilot, he'd transferred to the Royal Flying Corps during the last years of the Great War, which had ended before he had the chance to fly in combat. He stayed in the Service; in many ways he was the kind of man who found that the structure of military life suited his unstructured, not to say wild, temperament very well. Unfortunately his temperament didn't really suit the Air Force. Posted to India, he was rather too lax with his mess bills and spent more time playing polo than was wise; back in England he visited a touring rodeo in London which was offering a £10 prize for anyone who could stay aboard the mechanical bucking bronco for two minutes. In two weeks no one had managed more than half the time. Lancaster, who was always chronically short of cash – he was terminally generous with it – decided to take the challenge and, wearing a pinstripe suit and bowler hat, climbed aboard and, cheered on by his supporters, stayed in the saddle for the whole 120 seconds. It was quite a feat, but his superiors in the Service didn't see it like that. Nor did they approve of his participation in a parachute jumping exhibition in 1925, only weeks after the Air Force had finally allowed its pilots to use the things. Somehow, in almost everything the man did, there was a sort of insouciant snook-cocking, something you could never quite put your finger on but that marked him, in the official mind, as 'not the right sort of chap for us'. And he didn't really care.

He was full of confidence and the thought of consequences didn't exist in his mind. He was eccentric, naughty and very possibly dangerous to know. He was also married to Maud, called Kiki, and had two young daughters, but that wasn't important either. Kiki was working to support the family; Bill was planning his flight, and any money he was trying to raise was for Australia, rather than his wife and kiddies; not that he was very successful at it – the first time he met Chubbie, he touched her for a fiver!

He also told her something of his plans. She was fascinated, and the two spent most of the evening together, talking about flying, of which she knew nothing, and Australia, of which he knew something, since his family had spent time there when he was a boy. At the end of the party he invited her to tea at the Authors' Club and she accepted. He didn't mention his wife.

There would be 38 stopovers, Lancaster explained over the Darjeeling tea and cucumber sandwiches, or 39 counting Darwin, where he planned to land. He had worked out the distances and the fuel he would need and the plane he planned to use; the new Avro Avian would be perfect, light but with good range and capable of lifting almost its own weight. No one had done it in a small plane before; it would make the name of the first pilot to do so and he wanted that name to be Lancaster. As they talked, Chubbie became more and more excited by this man's enthusiasm. She asked when he was setting off and he shrugged and smiled that charming smile of his: he didn't know, it was the money, you see. He didn't have any. Not a bean. Even his father, who had plenty of money, wouldn't fork out – perhaps because of the family. The family? He explained how Kiki was working to support the children; he also hinted that there was no longer any love between them, no real relationship, that they were friends, that each went their own way . . . but that he desperately wanted to do the right thing and support them. And if he could get this flight going, then the money would surely pour in.

Chubbie had never flown, she had no interest in planes and certainly not in their engines but the idea of doing it, of *being* the news that Miller, back in Australia, was reporting, was something else: almost breathtaking. It would be a real adventure. And she said, right there and then, if I can raise the money, can I fly with you, as a passenger or a co-pilot or whatever? He thought about it, but not for long; he wasn't a thinking man, he preferred action, and this looked like the real thing. He said, you'll have to come as mechanic and clean the plugs and service the engine. And she said, fine. It says something about Bill Lancaster, who was, after all, 30 years old in 1928, that he listened to this absurd proposal from a woman in her early 20s, who had never stepped into a plane in her life, never mind left the ground in one, someone he'd known for four or five hours, and that he agreed to it.

Chubbie got to work at once. She walked round London and went to see every Australian company she could find and asked them for money. More often than not, they agreed. Australia had become federated in 1900, she had just gone through the horrors of the Great War in which she had played a more than creditable part and paid a high price in lives; the island continent was feeling its nationhood and the sight of a feisty and pretty young Australian woman proposing a world-record flight back home was 'bloody irresistible!'

For all their generosity, the businesses could not provide more than a small proportion of what was needed. Chubbie's next port of call was Sir Sydney Kidman, the outback cattle king, the richest, and meanest, man in Australia. (When asked why he always travelled second class on trains, he answered, 'Because there's no third class'.) He was also bowled over by her, and contributed handsomely. She even wrote to Miller back in Melbourne and got him to cough up, and then she went along to Fleet Street, to the most forward-looking popular newspaper of the time, Lord Beaverbrook's *Daily Express*, and offered them the exclusive story. They bought it, on condition that she telegraphed back at every stop – she insisted on reversing the charges – to keep their readers informed of her progress and of any adventures. They were almost there; but not quite. Chubbie cabled her mother to say she would be flying home soon. Mrs Beveridge, horrified, cabled back: don't fly, come by sea, and she sent a money transfer to cover the cost of the voyage. It went into the pot.

They found a cheap Avro Avian for £300, and Lancaster started to overhaul the Cirrus 111 engine. They replaced the wooden prop with a metal one they borrowed and went back to Avian to get extra fuel tanks installed. The fuel itself was donated by BP and Shell, and would be stored in drums at the appropriate stopping points. The only thing they couldn't afford was the insurance cover, and they hoped that Lancaster's father would, eventually, stump up for it.

There *was* one other thing, though; the elephant in the cockpit, so to speak. Mrs Kiki Lancaster. What was she going to say about her husband flying thousands of miles alone with a young woman? Chubbie knew what Bill had said but then she knew Bill by now and knew he would say almost anything, so long as it wasn't a big lie. He could never quite do that; his untruths were always little white lies. She told him they'd better go down to the country and see Kiki, tell her their plans and find out what she thought about it all.

Mrs Lancaster was running a hotel for vegetarians in Bournemouth, and Bill said he could arrange a lift down with a man called Robinson who had a car. Chubbie wasn't best pleased to discover it was an open-topped Morris Cowley, which meant she'd arrive, and face Kiki for the first time, with her hair a total mess. So she jammed on a hat and sat in the back while Lancaster and Robinson sat in the front chatting. They drove down so fast that she lost her hat and her hair got messed up anyway. However, Robinson obviously wasn't the motorist he thought he was; they got lost and didn't arrive until well after dark, so at least she didn't have to meet Mrs Lancaster that night. She was shown to a room and went to sleep at once. But Kiki was no fool. Chubbie was woken by her perfectly groomed hostess early the next morning, all gummy-eyed and bleary. It was, perhaps, all the victory Kiki wanted at the time, because she seemed quite nice and later the three of them went for a walk together on the Downs. Bill boyishly challenged the girls to a running race down the hillside. Kiki declined; Chubbie joined in. Afterwards, when she'd climbed back up and sat down beside her co-pilot's wife and lit a cigarette, she came straight out with the question: 'Now, look, you are quite sure you don't mind if I go with Bill on this flight?' Kiki said, 'Dear, I couldn't care less who he flies with or what he does, as long as he sends me money.' So that was settled.

Everything was ready, except for the insurance (Bill was going to ask his father) and the small fact that Chubbie had still never flown. What if she were airsick? What if she had vertigo? What if she hated it? They needed to be sure but had to be subtle: after all, with the *Daily Express* involved, Chubbie was beginning to become known by the public; neither they nor the media would be particularly impressed by the non-flying flyer. It would also be useful, Bill thought, if she could find out a few practical things about being in a plane, like not putting her foot through the fabric and learning how to strap herself in, when the cameras weren't there. He arranged for her to go up with him in a borrowed RAF plane and she loved it. From the second the wheels left the ground with that slight bump and the plane fell away into the sky she didn't experience a moment of anxiety; it felt like the right place to be, her element.

The night before they were due to leave, the Under-Secretary for Air, Sir Sefton Branker, gave a dinner in their honour. He was a long-time supporter of flying and knew many of the pilots, male and female, who'd been setting the long-distance records during the 1920s and he was able to give Chubbie one vital piece of advice: buy a light, easily washable evening dress and a pair of comfortable evening shoes, since the flyers were sure to be invited to some formal event, a reception or a dinner, wherever they landed and she'd feel a lot more comfortable if she had something to change into. Chubbie took his advice and rushed out first thing next morning to buy a black chiffon frock and a pair of satin slippers which weighed virtually nothing (just as well, since her weight allowance was only 7 lbs) and were to prove invaluable during the trip.

The next day Mr Lancaster was still reluctant to come up with the insurance money and with the start date looming, Chubbie and Bill tried to find it themselves or, at least, persuade some company to put it through on the nod. Unfortunately, neither of them, although they were becoming better known, was well enough known for the companies to believe in the possible success of their flight. The day passed; it looked as if they would have to give it up for the time being. Except that by now neither of them was in the mood to call the flight off; they fuelled the Avian, climbed aboard, Kiki Lancaster held up a daughter for Daddy to kiss, Mr and Mrs Lancaster shook hands and wished everybody luck and Bill told his father, quietly, that there was still no insurance, Mr L. said they could not possibly fly without it, and Bill turned the ignition switch and the plane began to roll forward. They took off at about 2.40 p.m. and headed towards the Kent coast, where they landed two hours later at Lympne airfield, from where Bill phoned his father and finally got him to agree to pay for and put in place the insurance.

They lifted off from Lympne at 4.00 p.m. and turned south over the Channel, heading for their first stopover at Abbeville. Lancaster was studying the folding maps trying to locate their landing strip so he called forward to Chubbie in the front cockpit: 'You take her.' She grabbed the dual control stick with both hands. The plane shot up into the dark evening clouds. Bill bellowed, 'Put her down, you clot!' She pushed the stick forward, gently this time, feeling the plane respond. They resumed level flight. Bill kept his head down over the maps. She put her feet on to the rudder bar and began, very gently, to push one and then the other, feeling the movement of the plane and checking against the compass as it turned. As it got darker she didn't seem to find any problems with keeping level; she just knew where the horizon was. After a while, she didn't really know how long,

Lancaster called forward that he was taking her now, since they were approaching the field. She let the stick go regretfully. She knew she was good at this; a natural, as Bill was to tell her later.

That night they were entertained by British Imperial Airways pilots in Paris and ended up dancing barefoot in the fountain outside the Folies-Bergère; the next morning they set off on time, though, in fact, Lancaster was not worried about keeping to a schedule: the two previous Britain–Australia flights had been by large or multiengined planes and in their small Avian they were setting the benchmark.

The first few days were easy flying and easier landing: across Europe, then over the Mediterranean, Morocco, Egypt, east to Baghdad and along the Persian Gulf to India, where, on the Calcutta stop, they received a bonus of several hundred pounds from the *Daily Express*, which was more than happy with the reports Chubbie was sending back. Lancaster insisted that he look after the money, though it had always been Chubbie who had arranged their finances. He said that since this was real currency, it might be stolen. In fact, the question had never arisen before since they'd never had any 'real currency', but keep it he did and lose it he did, leaving it under his pillow when they set off the next day. When they wired back to the hotel to see if it was still there, naturally it wasn't. Chubbie was furious. It was such a typical Bill Lancaster thing to do. But then, just because it *was* such a Bill Lancaster thing to do she found it impossible to stay angry at him for long because there was something about him she was beginning to find attractive. Perhaps it was a result of spending so much time alone with him; she was, after all a young, healthy woman, living life to the full every minute of the day and he was a handsome man and very decent even if . . . but then the situation didn't really allow for 'even if'. It was about absolutes, as they discovered the further they got from home.

Chubbie was flying the ship as they approached Rangoon. She was about to hand over to Bill, since he did the landings, not only because of her inexperience but also because the front (co-pilot) cockpit had an extra fuel tank in it and this made it impossible for Chubbie to push the stick far enough forward to bring them down. She was easing it forward, preparatory to handing over, when a bump caused it to jam under the instrument panel. It locked the dual-control system, so Bill couldn't take over and Chubbie couldn't pull the plane out of the shallow dive it was in. Down and down they went, with Bill bellowing at her to clear her stick. She tugged and tugged and the ground got closer and both braced themselves for a crash. Chubbie exerted every centimetre of her five-foot frame and hauled one last time. The stick came free, Bill grabbed control; they were approaching the Rangoon racecourse, it was flat and empty so he took her in, safely. As they rolled to a stop, an RAF truck arrived and the crew said they'd pull the Avian across to the local flying field while Chubbie and Bill cleaned up. The Air Force engineers advised the flyers to cover their cockpits as much as possible to guard against snakes crawling in overnight. The local kraits were particularly limber and likely to explore, and their bite was instantaneously fatal. Bill and Chubbie thought they were being made the victims of a joke, but they locked the craft down as well as they could.

That evening the two were entertained by the Royal Air Force squadron based locally; another outing for the invaluable black dress! Getting it washed and pressed was Chubbie's first order of business on landing; only then did she cable the *Daily Express*.

Refuelled and checked overnight by local engineers and checked again in the morning by Chubbie, who was getting a permanent oil stain on her stomach, where she pressed against the engine to reach the plugs, they took off for Tavoy. The day was good, the weather calm. Chubbie was flying when Bill reached forward to slap her on the helmet (the normal mode of attracting her attention against the engine noise). When she twisted back he pointed frantically down at the floor which ran, unimpeded, between front and back cockpits. She looked down and saw the unmistakable blunt-ended, white-bellied, black-backed shape of a krait. This time she had no trouble in relinquishing control. She ripped the stick from its housing and began wildly clubbing the floor, her feet and legs and, she hoped, the snake. At last blood began to spatter the cockpit walls and she relaxed. They were both unbitten but as soon as they landed that night they checked every inch of the ship just in case.

At Singapore they landed once again on a racecourse. After this stop they would be island-hopping down through the Dutch East Indies and then making the final leg across the Timor Sea, in all another 2500 miles. Since they weren't in a hurry they were happy to accept the hospitality of the local flying club and attend a banquet the following day. Muntok was the next stop, where they topped up fuel for the Batavia hop, taking off from the tiny island airstrip at seven in the morning. This allowed them at least some cool flying time before the sun began to beat down on them. The plane was heavy, but no heavier than usual, and the flying field sloped gently downhill. Lancaster pulled back on the stick and eased the flaps as they began to rise, reaching 150 feet before the engine cut out. Dead stick: no power, no speed, full fuel tanks, the ship was a falling incendiary bomb and they both knew it. Bill tried to sideslip and lose as much height as possible under some sort of control but the ground came up like a train and slammed into them, flipping the plane over on to its back.

Chubbie found herself hanging upside down by her knees from the flying wires that applied tension to the wings, while petrol poured down over her. She was wearing a sun helmet that morning, which probably saved her a fractured skull but now its rigid shape and size were pinning her head in the twisted wires and she couldn't get her hands up to unfasten it. Her eyes were smarting from the fumes of the fuel and, in the silence after the crash, she could hear the petrol hissing as it fell on the hot engine surfaces. She knew she had to get out before the plane caught fire and she began to wriggle, like a cork from a bottle, upwards through the wires and broken struts until she was clear. Lancaster was lying about 20 feet away and she crawled over to him. He was unconscious. Blood was pouring from his mouth where his teeth had ripped through his lower lip. She unhooked them and cleaned the blood from his mouth and tried to bring him round.

Dutch soldiers from the local garrison came running to help, and Bill was carried off to the local hospital. Chubbie was cleaned up and painted with iodine. Both her eyes were blackened and her nose was badly squashed. A peg was pushed in to get it back into alignment. Bill's lip was stitched and, once he'd come round, it was clear he had no other injuries beyond a bruised body. They did, however, have a smashed plane and, it must have seemed, irretrievably broken dreams. The Avian was disassembled and shipped aboard a junk back to Singapore, where it joined the recovering flyers. Back in England, when the news came through, the sponsors and Bill's family cabled: come home immediately. But it wasn't to be, not if Chubbie could help it. She cabled to Britain for spare parts and, as soon as they arrived, an RAF engineering team, who were travelling with an exhibition flight of flying-boats, volunteered to rebuild the craft. They had been so struck by a picture in the *Straits Times* of Chubbie looking like a rather cute panda with her two black eyes, and by her determination and courage, that they had no hesitation in giving up their free time to do the work. Besides, Bill Lancaster was an old RAF man, so they couldn't let him down either.

The cause of the crash had been Bill's inattention that morning at Muntok; he had forgotten to switch on the engine fuel lines and, starved of petrol, the engine had cut out. It was a mistake anyone might have made after so many miles in the air, but it was a mistake a great pilot should not have made.

Still, they had a good time in Singapore while the ship was rebuilt, and were fit and ready to start again on 17 March 1929, two months after the crash. This time Lancaster double-checked the switches before they left Muntok. They flew on down the archipelago, landing at Surabaya, Bima and, finally, Attambue, a Dutch prison colony. This would be the starting point for the last leg, the 500-mile Timor Sea crossing to Darwin. They would need to fill every fuel tank to the brim but this would increase the dead weight of the plane and the landing strip at Attambue was soggy; the plane would bog down as it tried to taxi. It seemed an impossible dilemma: take off without enough fuel; load enough fuel and stay on the ground. They asked the governor of the local prison if there was anything he could think of – no problem; he had his shackled prisoners shambling out in no time to lay coconut matting all over the strip, layer after layer, until it was dry and supportive enough for the fully-loaded Avian.

They started early, heading out over the unruffled sea, flying level for hour after hour, covering an almost hypnotic progression of miles until the engine started popping and they came instantly to their senses. Each time the engine stuttered they lost power and height, sinking down towards the ocean surface and whatever waited for them just beneath it. Then the motor would catch and Lancaster would gain height, and then it would start to miss and down they would go, closer and closer to the water each time. There was no wind; the surface was like copper and would be as hard as steel, they knew, if they hit it at the wrong angle. They didn't carry an inflatable dinghy; flyers crossing the Timor Sea never bothered with such things; if they went down they didn't expect to survive. Bill struggled on; the engine caught and failed. He wrote a note and handed it to Chubbie: 'I don't think we're going to make it, old girl, but it's been a damn good try.'

She reached back and they took hold of each other's hands for a moment, then let go and sat quietly, Bill flying for as long as the engine held any power at all, Chubbie just waiting. They both knew what would happen when the ship went down.

The motor spluttered and stopped, the stick dead, the prop spinning slowly in the wind, all around them was silence, broken only by the hum of the wing wires cutting through the wind; and then, for no ascertainable reason, the engine took hold and began to turn with all its old steadiness. The plane pulled up, clear of the water. They were through the worst; they knew they would make it now. And they did, coming in to land at Darwin in about a foot of water. The airstrip had been flooded and the authorities had cabled Attembue that on no account were the flyers to come that day. But now here they were and a solitary airport employee stood shouting at them from a hundred yards away that he could not come any closer until they'd been checked for any tropical diseases they might have imported.

At length a doctor did arrive and they were able to show him the certificates for the hundreds of jabs they'd collected on their journey, which was not, now, the first Britain–Australia flight by a light plane. The aviator Bert Hinkler had started from London on 7 February, after their crash, and had used their enforced layover in Singapore to arrive first. Lancaster, ever honourable, had been adamant that their Singapore supporters, who had hinted that Hinkler's bad sportsmanship would not win him the first prize, should do nothing to sabotage their rival's craft. He even spent one night in Hinkler's cockpit, while the aviator slept, to ensure that no dirty tricks were pulled.

Although their triumph was not the first, Chubbie's presence ensured that, for Australians, it was certainly the best! Bill Lancaster's father, who had sailed out while the flyers were stuck in Singapore, tried to persuade his son to drop Chubbie from the tour and receptions; Bill, naturally, refused. Besides, as he pointed out, the crowds weren't really interested in him, it was Chubbie they wanted to see and she was the one who was offered the lecture tour. Bill flew her between engagements. The couple were quite happy with each other's company; Chubbie pointedly didn't see Miller who was still her husband (although the two did agree by letter to start divorce proceedings). She had also agreed with Bill that a third of their joint earnings (in fact, her earnings, since she was making most of the money) should go to Bill's wife Kiki. It seemed like a good idea at the time; there was no shortage of work and the country was in a state of flying euphoria. Chubbie and Bill were present in Brisbane when Australian pilots Charles Kingsford-Smith and Charles Ulm with US navigator Harry Lyon and wireless operator James Warner completed the first flight across the Pacific.

At a reception for the four flyers, Chubbie and Bill got to know Harry Lyon, a larger than life character who suggested the three of them should get together and set up an Atlantic crossing, possibly to be followed by an Australia–America flight. It sounded like a good idea to Chubbie, who hadn't really given any thought to her personal situation, other than acknowledging the fact that she and Bill were lovers but not in love. Bill, plagued by his strictly conventional father to return home to wife and family, found the idea of America even more attractive; they accepted Harry Lyon's proposal and sailed with him and radio operator Warner back to San Francisco, where the city set out to honour the two American heroes with a tickertape parade and, with typical generosity, included Chubbie and Bill as well. They rode through cheering crowds to a reception at City Hall. Before the big public banquet all four were given a few minutes to relax in a side room. The mayor asked Chubbie what she'd like to drink; something to pep her up before the public ordeal, maybe? She asked, wasn't Prohibition still in force? The room erupted with howls of mirth, and the mayor pulled aside some sliding doors to show a whole wall lined with bottles, glasses and ice boxes!

While Bill looked for work, Chubbie got in touch with a lecture agency and arranged for a tour of the country; male pilots, even handsome charming ones like Bill Lancaster, were two a penny; women were still something of a novelty, particularly if they could fly into town, give their show, and fly out again. There was only one problem as far as Chubbie was concerned: she didn't have a licence. All her flying had been done with Bill; she'd never taken an official lesson in her life. She slipped away to New Jersey for a few weeks and an intensive series of lessons, and then took and passed her test and received the vital licence. She could now join the lecture tour which had set out earlier, since she was only part of the package, appearing alongside a small orchestra and a comedian.

Meanwhile, plans were going ahead for the Atlantic flight but with increasing difficulty as it became clear that Harry Lyon knew which sliding doors to open in every building he ever visited, and spent most of his time either drunk or lost, not a good augury for an Atlantic navigator. The backers of the flight were beginning to hint that Chubbie and Bill should go alone, which meant that Chubbie, as co-pilot, would need to take care of radio communication. She knew nothing about wireless and went off to take a course in the subject, a fortunate absence since Kiki Lancaster had cabled that she was fed up with sitting alone and poor in England while Bill got all the glory in America, and she was arriving by the next liner. At the time, Bill was working as a representative for Cirrus engines in Chicago, where he got to know Lady Mary Heath well enough for her to take Kiki (with whom she sympathised) off his hands for a flight to Miami. By the time Chubbie had finished her course, Kiki had departed for England.

Chubbie and Bill could be together with Kiki out of the country but the material question meant that this was, in fact, far from easy. Cirrus, a British based firm, had decided to start building their engines from scratch in the States and to celebrate this, with the maximum amount of publicity, they entered Bill into a big-money New York–Trinidad and back race. He would be flying an Avro Avian, a plane he knew well, with the new US Cirrus engine installed; it was a chance for the company to show the qualities of the engineering and for Bill to show his skills at solo long distance flying. Unfortunately for both, he crashed on landing in Venezuela and was badly injured, spending three months, and money the couple could ill afford, in a local hospital before returning to New York for another two months recuperation.

Chubbie was looking round for something of her own and settled on the 1929 Powder Puff Derby; she reckoned she had just enough experience behind her now, though most of her air miles were borrowed from the London–Darwin flight. This wasn't so much of a problem as the lack of a plane. She couldn't afford to buy anything that would be good enough for the race, so she went to see Laurence Bell at Bell Aircraft and persuaded him to lend her a Fleet biplane with a Kinner K-5 engine; she must have used all her considerable will and a touch of charm as well since Bell agreed not only to modify the plane for racing but also to rebuild the cockpit around Chubbie's five-foot height. She had to test-fly the craft at the Bell field, something of an ordeal since she had, in effect, only just gained her licence and now would have to take off and land in front of a highly critical audience of pilots and engineers. Any mess-up could easily mean the withdrawal of the plane: no company wants its name associated with an amateur! However, her natural ability got her through and she flew down to California in time for the start of the race. She and Thea Rausch from Germany were the only foreign pilots. Lady Mary Heath had entered but her business commitments kept her otherwise engaged, though she would be racing at the Cleveland air show that marked the end of the Derby.

Chubbie's race went smoothly from the start. She was among the leaders at the first stop and found herself sharing a room and the inevitable chicken dinner presented by every town along the route, with Amelia Earhart. The two got on well enough, though Chubbie found her roommate's reserve slightly disconcerting. However, on the second day, after Marvel Crossen's fatal crash, Chubbie and Earhart were among the most vocal in their support of the race continuing.

The Fleet Kinner was one of the lighter planes competing and Chubbie was beginning to feel she had a real chance of placing first in her class. However, on the penultimate day she found herself with engine trouble and was forced to land in a field, coming to a stop only inches from the edge of a deep pit the rancher had been digging. A quick check revealed a cracked cylinder, and the Fleet engineers waiting at the finish were cabled to fly one down as fast as they could. They got to it; the part was duly delivered and fitted, and Chubbie was back in the air fast enough to be able to claim third place in the light aircraft section, behind Phoebie Omlie and Edith Foltz. She was awarded a cup, and feeling that things were surely going her way, entered three pylon races at the Cleveland meeting. She achieved a first, second and third and, apart from receiving a big bonus from the delighted Bell Aircraft Company, got some extremely useful publicity.

On the strength of this, Fairchild Aircraft approached her about racing for them. She was happy to do so and they entered her into the 3000 mile Ford Reliability Race. They also persuaded her to dress entirely in black and white: white silk shirt and jodhpurs, white jacket and helmet, black boots and tie. Her new plane, too, was painted black and white. With her (from an American point of view) exotic accent and diminutive size, she made a striking, if small, figure and attracted considerable attention as one of only three women entrants in the race. When the other two dropped out and she placed eighth, her reputation as a gutsy endurance flyer was firmly established. Employers and sponsors were eager to give her as much support as she needed for any future races or record attempts. Which was just as well since Bill wasn't earning and had lost his job with Cirrus after the crash and though he was fit to fly, by the beginning of 1930, the Depression was closing in and there were too many male pilots chasing too few jobs. For once, being a woman in the air was an advantage; Chubbie was still a novelty and she was picking up advertising stints and making distance flights across the continent and earning enough for both. She had a flat in New York; Bill stayed at the Army and Navy Club: they didn't live together, partly because of the fear of bad publicity (Chubbie was known as Mrs Keith Miller in America) and partly because she didn't want to. While the couple had been apart, Chubbie had been quite content with her own company, though she had continued to pay an allowance to Kiki and the children; Bill, on the other hand, had been as lovelorn as a teenager. She was beginning to feel there was something a little overheated about his dependence on her; it was almost as if she was taking the places of his wife Kiki *and* his mother Maud, a proselytising Christian spiritualist known as Sister Red Rose. And she didn't know that she liked the idea.

However, she was still extremely fond of Bill and when the chance of a job together came up, she was happy to go along with it. At a party (obviously parties were lucky for Chubbie) they met C. T. Stork, the chairman of an aero company which had a concession to sell a number of flight-related products as well as engines and planes. He offered them jobs as display pilots at $600 a month each. It was an offer they couldn't refuse. They accepted at once. Stork asked, as an afterthought, that, of course, they did both possess US commercial licences? Chubbie felt a sudden chill but, before she could say they hadn't, Bill jumped in and told the Dutchman that, yes, of course they did. They agreed a starting date the following week, signed the contract and left. In the lift going down, Chubbie said: 'We have no licences, Bill.' He said, 'But we are going to have them.' He reminded her of an old friend, J. R. Boothe, who ran a flying club in Ottawa, and went on, 'He said any time we want any help to call on him. I'm going to send him a telegram now, and say we'll be catching a train this evening, will he meet us in the morning and fix with the Canadian Air Force to let us have our tests and to give us commercial licences.'

That's what they did – got on a train and headed north, with Chubbie, who was dressed for the city, getting colder and colder by the mile. When they reached Ottawa, Booth was waiting for them on the platform with a full-length fur coat for each and an offer of lunch before they met the medical examiner and then took their tests. Booth was a convivial sort, as were Bill and Chubbie, and though she didn't drink a lot, with her small frame, a little went a long way and by the time she arrived at the medical examiner that afternoon, her eyes were rolling and the eyesight test proved impossible. The doctor was puzzled: she should be able to pass, what was the problem? Big lunch at the Silver Slipper, she told him. He was obviously a gourmet himself, since he understood and told her to take the flying tests, then come back for the sight test when her head was clearer. They both passed with ease, got their licences and stepped back aboard the evening train for the 16-hour return journey. It isn't known whether or not Chubbie kept the coat.

The job with the Stork Company lasted six months; among their concessions was one to sell Stinson planes, and Chubbie, while taking the company course, was praised by Eddie Stinson, pioneer builder, pilot and brother to Marjory and Katherine, who told her she flew as well as any man he'd ever seen.

She got to fly a whole range of ships during her time with Stork, among them the Savoia Marchetti, an Italian float-plane which could also take off from land. She was demonstrating it to a rich young man who arrived in the most beautiful Deusenberg she'd ever seen. Unfortunately, she'd never seen the Marchetti before and had to fly it by instinct. Luckily, her instinct was good but once they were airborne, she realised that the pilot had to pump the wheels up by working a metal lever. There were no hydraulics; it was all muscle and weight, and at just under seven stone, Chubbie had neither. She told the customer she'd sprained her wrist recently, could he help? He was glad to do so. The ship itself was a delight to fly, though as they came in to land on the glassy surface of the lake next to the test field, Chubbie realised that she'd never done this before either. The most she remembered was hearing somewhere that you should always create a few ripples before touching down or your floats might crumple. For once there seemed nothing in the cockpit, not a spanner or spare spark-plug she could throw out; in the end she went through her bag and dropped her compact. It did the trick and she landed perfectly. The customer was delighted and said he would place an order at once, although she never found out if he did, since their time with the company was up. Everyone was pulling back in the face of the Depression and Stork was no different than any other businessman.

They were out of a job again, but this time they had enough money on hand to make plans for the future. Bill, acting as Chubbie's manager, since she was once again more saleable than he was, met a representative from Aerial Enterprises, one of the few air transport companies still spending money on advertising stunts. They were offering $1000 for a flight from Pittsburgh to Havana and back, to deliver an illuminated scroll from the mayor of the city to the Cuban President. It wasn't a long distance, although the weather over Miami and the Gulf of Mexico could be tricky, but it would keep Chubbie in the public eye and publicity was more important than anything else in these lean times. The contract was signed and Chubbie took off from Pittsburgh, complete with illuminated scroll, on 22 November. She was flying an Alexander Bullet, a low-wing monoplane which was still in development, mainly because it had a tendency to spin out of control even in the hands of an experienced pilot. It was, however, cheap and available – and Aerial Enterprises weren't providing a plane – and Chubbie had no intention of doing aerobatics in the thing anyway; she'd be flying straight and level, and in those conditions, the ship handled perfectly.

On the trip down, the weather was rough but not too rough. She reached Cuba with no problems, delivered the scroll and was wined and dined overnight at the British embassy. The next morning, however, the weather over the Gulf had closed in, conditions were bad and she cabled Pittsburgh, saying she would postpone take-off until they improved. Aerial Enterprises were not happy; the President of Cuba was sending his own address back to the mayor of Pittsburgh, who would be waiting to receive it and all the attendant publicity. Chubbie was urged to see the vital importance of not keeping the mayor and the media waiting. Despite her forebodings, she did as she was asked, and set off on the return journey.

The weather was bad over Cuba but once she got clear of the land, it was even worse. Her tanks weren't full, she'd decided to fly light and refuel in Miami before going to

Pittsburgh, but something made her return to Havana Airport and fill the Bullet's tanks to the brim before starting back again.

It was just as well she had; the clouds came down at once, she was flying on instruments and then she wasn't flying at all, the winds were doing it for her; the Bullet was simply snatched from its course and thrown 300 miles out over the ocean. She was still flying virtually blind, the wind shrieking, the plane being thrown around the sky in demented aerobatics so that every second she expected it to flip into its notorious spin and smash into the waves. She had no idea how long she battled the storm or how far out of her course she'd been taken, but eventually the wind speed began to fall and she could make out rifts of sea through the clouds, and then the waves began to settle as she flew on and on. The main tanks were empty; the engine tanks would last only a hundred miles or so, so she scanned the horizon for the coast of Florida. At last, she saw green water and then land.

She began looking for a town, some familiar landmark, an airstrip, perhaps, but could see nothing, not even a village. Maybe this wasn't the Florida Keys at all. Should she try and land, or climb in the hope of making out exactly where she had ended up? The decision was taken out of her hands when the engine began to splutter. She was out of fuel. Below her there was a beach and she circled down towards it. The sand seemed clear and she came in to land, damaging her wings on some coarse bushes. She sat in the cockpit breathing deeply, realising just how close she'd been to dying out there over the ocean; if she hadn't gone back to fill her tanks, that would have been the end.

As far as the rest of the country was concerned, that was just what had happened. The newspapers headlined the search for the missing Aviatrix: NO MORE THAN A THOUSAND TO ONE CHANCE OF A SAFE LANDING. Planes and ships in the Gulf were asked to look out for the missing flyer, but when two full days resulted in not a single sighting, the authorities began to give up hope. It wasn't unreasonable. If Chubbie had come down and hit the sea during the storm she wouldn't have survived more than a couple of minutes; if she'd managed to get back over land, then why was there no word? Only Bill Lancaster, a man never in the closest touch with reality, refused to believe it: she was out there somewhere, he maintained, and they must keep looking; but nobody was listening.

Chubbie clambered stiffly out of the Bullet's cabin and looked around her. The beach seemed deserted, then a group of men appeared from the trees. They looked like fishermen; she hoped they weren't pirates or smugglers, who were known to use the islands of the Gulf for their operations. She called across the sand and asked where she was. Andross Island, she was told. And who owns it? she said. King George. So she was in the Bahamas. The fishermen, for such they were, gave her a hand tying down her plane, shared some of their food (it seemed like an age since she'd eaten) and one of them volunteered to show her the way to the north island where there was a British residency.

She set off along the sand with her guide; the wind was still strong, whipping her hair across her face so violently she was getting bruised. Trees had been blown down and in places they had to climb over landslips or walk in the water. Night had fallen and Chubbie was reaching the end of her endurance. At last they found a beach bar, closed and shuttered against the storm; her guide told her it belonged to his brother-in-law and they stopped to get a lantern, the better to see their way. Chubbie was desperate for a drink, but there was no water in the place, only bacardi or gin, so she sank a few of each, which warmed her up, and set off again, following the lantern, until they reached another building, this one larger, her guide's own house, where Chubbie fell into bed with his wife and six children and slept the sleep of one who knew she had just flown wingtip to wingtip with death.

In the morning she moved along the island to a much larger house owned by a man called Forsyth who had a radio, which had been broken a few weeks before and which he hadn't bothered to get mended. His wife, however, did help Chubbie dress her feet, which were in a terrible condition, and Mr Forsyth mentioned that there was a famous Australian swimmer living on the island who might help her. The swimmer was summoned and arrived that night, just as Chubbie was abut to go to bed in a nightdress borrowed from her hostess. Mr Forsyth, a gentleman of the old school, refused to allow the visitor to see a lady in her night attire. She dressed hurriedly and the champion swimmer told her he had a sailing boat at the beach and could take her to Nassau. Once again Mr Forsyth wasn't keen: the sea was too rough; he might get blamed for letting the lady take a risk; besides, there was no chaperone for the voyage. In the end, though, Chubbie and her countryman managed to slip away and sail, overnight, to the Bahamian capital.

Bill was overjoyed – and so was the British ambassador in Washington, who would no longer be pestered day and night by this mad ex-RAF flying officer who appeared to have lost the love of his life somewhere over the Gulf of Mexico or maybe in the Everglades. Lancaster flew out to Nassau, bringing materials to mend the Bullet. The couple hired a float-plane and flew down the coast, where the local fishermen helped, rowing cans of fuel ashore. The Bullet was patched up and the pair flew her back to Miami. Once more, Chubbie was the centre of attention, giving interviews over the air and to the papers. However, she still had to deliver the address to the mayor of Pittsburgh and claim her $1000, so she set off again to make the delivery. As she left the ground, the engine failed. She was heading straight for a fence and row of trees at the end of the runway and, although it was against all the rules after losing power, she attempted to turn the plane. It did a ground loop, spinning on to its back. Chubbie was injured, not seriously but badly enough to stop the flight and give Aerial Enterprises an excuse not to award her the $1000 prize. It was simply a piece of bad luck.

Shortly after this, rumours began to surface in the newspapers that the whole Cuba flight and the forced landing was no more than a publicity stunt. Bill was stung by the insult to Chubbie and lambasted reporters, informing them that anyone who didn't believe 'this little lady' was not worth talking to. He proceeded to tell them to get out of the room. A grand gesture and, for a couple who depended on good publicity, an incredibly maladroit one.

The next few months were not good for either of them. Chubbie set up a number of flights but for one reason or another none of them came together. Lancaster found a position as personal assistant to the President of Intercontinental Airlines, but his management skills were simply not up to the job and he soon found himself unemployed again, looking for something, anything. What turned up was an old friend, Gentry Sheldon, who was still in the flying game and wanted to head to Miami to see the air races that were coming up. It seemed like the kind of place Chubbie and Bill might well make some good contacts and perhaps call in a few favours; they had ideas of starting a small charter business of their own, flying around the Gulf of Mexico. Bill had been given a car as a leaving bonus by Intercontinental so he suggested that he and Chubbie should drive Sheldon down to Florida. They all piled into the car and drove down to watch the air races and met a few old friends and made a few new ones.

Among them were a couple of adventurers called Tancrel and Russell, who called themselves Latin-American Airways and said they were thinking of setting up an air charter service and that maybe Bill and Chubbie could fly for them, if they got hold of their own planes. Tancrel and Russell would supply the funding and set up the routes between the US and Mexico and other, unspecified points south. It seemed like a reasonable idea to Bill, though he wasn't sure about Tancrel's credentials. The man appeared to know more about paper hanging, as in decorating, than running an airline. However, money was short and Bill promised to look seriously at the proposals once he got back from a trip east, where he'd been engaged to carry out a short-term charter. In the meantime, Chubbie would look around for a property to rent locally, so they could be on the spot for any developments. Chubbie said that she might also look around for a collaborator to work on the story of her life. Flyers like Amelia Earhart made big money from their books, so why shouldn't Chubbie cash in too? After all, no one could say she didn't have some good stories to tell.

Coral Gables, 2321 SW, 21st Terrace was a spacious two-storey house with its own orchard where grapefruit and lemons grew. It had a garage and extra parking. Inside, the rooms were large and cool. On the ground floor there was a big central room leading into a small dining room and a kitchen at the back; there was also a bedroom and a room for the icebox. At the top of the wide, open stairway was a landing with a bathroom off it, a bedroom which Chubbie took for herself and a large open area or sleep-out, as big as the central room and dining room together. There was no glass in the windows, just netting to keep out insects. There were five beds in this room, as well as closets and a couple of tables. Chubbie thought it would be ideal for their many flying friends who came through Miami to visit, since most of them were out of work and hard up too, or perhaps it could be rented out as a virtually self-contained flat.

After settling in, Chubbie rented the downstairs room to a flying friend, though 'rented' was something of a misnomer since he never managed to pay any rent and she was forced to ask him to leave after a while. That was a problem; none of their friends seemed to be in work, although most of them did seem to be in Miami, passing through and staying over for a few days, eating their way through the larder and offering no payment at all, either in cash or in kind. Times were hard, but they were as hard for Chubbie as anyone else, and she was running low on her savings and still paying out for Kiki Lancaster and her kids back in England. She began to feel just the tiniest bit resentful at all these men she seemed to be supporting.

As something of a local celebrity, Chubbie was invited to civic and commercial flying functions and at one of these she met a local matron, Mrs Helen Clarke. The two hit it off and Mrs Clarke invited Chubbie to tea. Remembering her first meeting with Bill, she might have thought this an auspicious portent; it wasn't, history *was* about to repeat itself, but as tragedy rather than comedy. Over the white linen and fine china, Chubbie happened to mention that she was looking for a writer to help her put together her autobiography. Mrs Clarke replied that her son Haden happened to be a writer who didn't have a project at the moment; perhaps he and Chubbie should meet? It sounded like a good idea, and a few days later Chubbie first laid eyes on Haden Clarke.

He was young, he told her he was 30, her age too, he was tall and dark with wavy hair, handsome and aware of it and charming too; she didn't really like him but then, as they began to talk, she found his enthusiasm for the project impressive and he was quite happy to work on the book with her without any kind of advance, just food and lodging and, after all, she had the big open room, so why shouldn't he use it as bedroom and workroom? But all the same, there *was* something . . . she couldn't put her finger on it . . . something about the man.

He moved into the big room at about the time Bill came back from his job opportunity in the east. Like so many of his projects, it had come to nothing. It wasn't his fault; Eddie Stinson had just died in an air crash and his passenger had got a case of the nerves and decided to do his travelling by rail. Bill wasn't cast down, though; it looked like Tancrel and Russell were getting Latin-American Airways off the ground. Messages were coming in from Mexico that things were looking good and Bill should get ready to go on west to discuss taking up his position as chief pilot. However, Bill Lancaster's return home put a stop to work on Chubbie's autobiography, not that much seemed to have been done so far. Haden spent a lot of his time huddled over the typewriter with a cigarette and a drink, and not much of it pounding the keys. He said he was a slow starter and would soon get all the inspiration he needed. Chubbie had always been more of a perspiration woman but, who could tell, writers were funny people and, when he set his mind to it, Haden was very pleasant to have around. Bill agreed. It didn't take long for the two men to become good friends. Perhaps, Chubbie felt, there was something fatherly in Bill's attitude, since though he was only a few years older than Haden, he seemed so much more mature, for all his boyish enthusiasms. He was also becoming possessive of Chubbie, never letting her go out alone when he was around, always placing himself between her and any man she met at a party.

There had been a time a year before, after she'd competed so successfully in the Ford Reliability Trial, that her sponsors Fairchild Aircraft had approached her with an offer of an executive job; however, it was on condition that she get rid of Lancaster. Perhaps it was the scandal attached to his name because of his desertion of his wife and children, perhaps something about the relationship between himself and Chubbie, but either way the company was fighting shy of the man. Chubbie, naturally, did not take up the offer but now, with Bill back in town without any money, still being supported by her, as was Haden Clark and any other male guest who stayed over, she began to wonder if it had been the right decision. Maybe she'd be better off on her own.

Bill still had hopes of gaining a divorce from Kiki; he was trying particularly hard to persuade her now that Chubbie's divorce from Miller had been finalised. He broached the subject in letters back to England but Kiki was adamant: without a good financial settlement in her hand (she knew Bill that well) of at least £10 000 there would be no chance. Chubbie wasn't too concerned; marriage, although she and Bill were still occasional lovers, was not part of her long-term plan. In the short term, things could go on as they were, if only she could get Haden to start work.

The writer seemed to be having more fun with Bill; the two would spend hours talking together or go out drinking or try to steal a chicken or two for the family pot – not that they had any electricity to cook it with, the company had cut them off for non-payment of bills. Chubbie was getting more and more frustrated with the lack of progress; even when she told him things, Haden appeared to make no notes of her words, he just lit up another joint (Chubbie did not approve of marijuana and hated being around when everyone else smoked it and started giggling at pointless jokes and getting hungry and cleaning out her larder and fridge *again*) and kept up with the same old refrain: the inspiration will come and then we'll really take off.

Bill, at least, looked like he might be taking off for once. Latin-American Airways were getting impatient; why hadn't he come down to Mexico yet? The situation got so tense that Tancrel made a journey east specifically to fetch his chief pilot. Bill refused to go until he had some money to leave Chubbie. Tancrel handed over $25 and promised a salary of $100 per week once Bill started flying for them. This was too good to refuse, and Bill said his goodbyes, but he didn't leave until he'd had a man-to-man talk with Haden. He asked his new friend to look after Chubbie, because, as he made quite clear, she was the most important thing in his life and, more than anything else, he wanted to look after her. Haden promised, in a manly sort of way, to keep his eye on Chubbie; Bill would have nothing to worry about.

Whether Haden made his promise in good faith or not, we don't know; he probably didn't know himself, but once his eye had been directed at Chubbie, with Bill's words ringing in his ears, his intentions became clear. He may not have been much of a writer, but he was one hell of a seducer. There were two or three women who used to hang around the house without Chubbie's knowledge, at least one of them staying overnight. The fact that he had venereal disease didn't seem to stop him. Neither was Bill's usual manner with Chubbie much of a barrier. His back-slapping 'jolly good old girl' 'little lady' style was not calculated to charm (a quality the English middle class has never trusted). And while Chubbie was no pushover, it is very difficult to resist someone you find attractive, who finds you utterly ravishing in return. Never once in their relationship had Bill even noticed what she wore; Haden not only noticed, but he made suggestions, had preferences, paid her compliments and always seemed pleased or excited to be in her company. And then there was Florida itself: it was a very romantic place with the smell of jasmine and a huge moon reflected in the water. One night there were kisses and Chubbie decided she must be in love with him; he asked her to marry him and she said yes.

He was certainly intelligent; his mother was a nice woman, though he didn't get on with her; he was artistic, charming... and actually, if you thought about it, the list stopped there. But Chubbie didn't think about it. Not really. It was a moment of emotional release; she'd been through a huge number of experiences in a few years, she'd been world-famous and now she was just another woman of 30, not getting any younger, with the prospect of Bill Lancaster spending the rest of his life never getting himself together. If Haden Clarke was a liar, at least he did it in style and, if you squinted hard enough in the moonlight, you could believe him for a while.

Over in Mexico Bill thought things were going badly. Latin-American Airways appeared even shiftier up close than they had from Miami. It was becoming clear that the passengers they envisaged carrying were illegal immigrants and the cargo illegal substances. Bill wrote a letter to Chubbie and Haden, asking their advice. Should he take up the job offer, knowing he would be breaking the law? He wanted, above all, to provide some money for Chubbie, but what would she think about ill-gotten gains such as these?

Chubbie and Haden wrote a letter back to Bill telling him that they were in love and planned to marry.

Bill cabled back: 'Hold your horses, I'm coming back, I want to be best man.' He began to look for a plane to borrow.

In Florida, Haden told Chubbie that although he wanted to marry as soon as possible there was the little matter of his venereal disease, which he thought had been cured but wasn't. He hoped she wouldn't mind. She did; she was furious. She thought she'd never get over it but then, she loved the man, he had done the decent thing and confessed; they hadn't slept together so she would be understanding about the whole business. Haden really appreciated this (he was a man who had relied upon the understanding of many people in his life) and suggested they marry anyway and just hold off physical relations. He didn't mention that his own divorce had not come through yet, or indeed, that he had been married before. Chubbie realised that he wanted to get the business done before Bill got back but she was sure of her own feelings (or she thought she was sure) and she knew that Haden would be steadfast, and anyway, they had to play fair with good old Bill.

Who was buying a .38 Colt revolver in St Louis at that moment. He had borrowed one from a friend before going to Mexico and lost it and wanted to be sure of replacing it when he returned. He flew on from St Louis and landed that evening in Nashville, then hopped to Atlanta, from where he cabled the happy couple to expect him in Miami by half past four that afternoon.

They were at the airport to meet him. Everyone was scrupulously polite. During the drive back to the house, Bill asked Chubbie if she was absolutely sure that she was doing the right thing. He then asked Haden if he thought he could make Chubbie happy. Yes, absolutely, the writer assured him. Bill then asked the couple to wait a month, just to be sure, and told them that on their wedding day he'd give them a cheque for $1000. Chubbie had no idea where he'd got the money from, but she was relieved that things seemed to be going fairly well so far. Back at the house she cooked dinner, which they ate in a 'rather strained atmosphere', and, at last, Bill said, 'What's this nonsense you tell me?'

Not nonsense, they both assured him. He looked straight at Haden and said, 'You've let me down. You haven't behaved like a gentleman.' Haden was furious and leaped up, ready to defend his honour. Chubbie didn't want to sit and watch them arguing, so she told them she was going to bed, which she did, taking her wire-haired terrier with her and locking her door. There didn't appear to be any noise from downstairs or from the big room, where both men would have to sleep, unless Bill took himself off, which was the most likely thing to happen. She read for a while, then turned off the light and went to sleep.

She woke up some time later. Bill was hammering on her door, bellowing: 'Chubbie, Chubbie, get up quick for God's sake, Haden's shot himself.'

She got out of bed and ran to the door, opening it to see Bill standing there on the landing, wild-eyed and wild-haired. She told him not to be a damned fool, there wasn't a gun in the house, how could Haden have shot himself? She thought it was a black, black joke.

'Oh yes there is,' Bill said, 'I brought one with me.'

She pulled on her dressing gown and followed Lancaster into the big room where they found Haden lying on his bed groaning. There was a hole in his temple and blood on the pillow. She couldn't see a gun, and Bill pulled the sheet down: it was there, half-hidden under Haden's body. She went to pull it out but Bill stopped her, telling her she might get fingerprints on it. He then pointed out two notes Haden had written on the typewriter. One was addressed to Bill and said, 'I can't make the grade, tell Chubbie of our talk. My advice is, never leave her again.' The second was to Chubbie herself: 'The economic situation is such I can't go through with it. Comfort Mother in her sorrow. You have Bill. He is the whitest man I know.'

Bill thought they'd better destroy them both; Chubbie disagreed, they were evidence. Ever practical, she hurried downstairs to call a doctor but when she got through and explained what had happened he told her to call the police, which she did. She stood in the hall and waited for them. When they arrived, she went back upstairs with them. Haden was still groaning and she was hoping desperately that it wasn't fatal and trying to work out what had happened. An ambulance had arrived and also a solicitor friend, Ernest Huston. It had been Huston from whom Bill had borrowed the lost pistol and he took the lawyer aside to ask if he could say that the new .38 he'd brought back, and which Haden had used to shoot himself, was, technically at least, Huston's. Absolutely not, the lawyer said; it just wouldn't do. However, as Bill and Chubbie were taken to the Dade County Courthouse to be questioned about the incident, he did agree to act for them, should it become necessary.

At the courthouse two detectives went through the events of the night with Chubbie and Bill. She was allowed to go home but, since Bill had been in the same room as the now deceased Haden – he had died on arrival at the hospital – he was detained while the police went through his letters and diaries, where most of the events in the week or so leading up to the incident were detailed. There was also a question about the notes, neither of which matched examples of Haden's spelling and typing; both the notes looked more like Lancaster's.

Later that same day Chubbie attended Haden's funeral with his mother, then returned to the courthouse for more questioning. The state attorney, N. Vernon Hawthorne, told her he was quite satisfied with her version of events and she was free to return home. Bill Lancaster would be held for a while yet, until various questions were cleared up; however, after reading Bill's diaries, Hawthorne released him too, since he could find no evidence of malevolence; indeed, he thought the diarist one of the most honourable men he'd ever encountered. As for the suicide notes, there was absolutely no proof that Haden hadn't written them.

Once the couple arrived home, Bill told Chubbie that he had written both the notes after being woken up by the gunshot. He said that he'd written them solely for Chubbie, so she wouldn't believe he'd killed her fiancé; that was the reason he'd wanted to destroy them. Chubbie asked him if he had killed Haden. She said, 'Bill, did you do this?' He replied, 'On my honour, I did not.'

She knew him better than any other human being, even his mother and father. They had never understood the kind of man he was, any more than his wife had. Chubbie knew; she had flown with him, faced death with him, seen him at his worst and at his best; she had no illusions about him, he had just been deeply involved, to say the least, with the death of a man she was going to marry; but she was a woman who could keep cool in a hurricane over the ocean; she had great intelligence and clarity of vision; she was a pilot; he was a pilot and she believed him.

She told him so, then called for a taxi to take her back to the police station so she could tell them about the notes. Better from her now, than later from a prosecutor. Bill said that Haden's supposed words ending the note to Chubbie, 'He is the whitest man I know', were, in fact, the last words Haden had spoken to Lancaster before they settled down to sleep.

After she had told the police about the notes, Chubbie was questioned again and then allowed to leave. Bill was called in and questioned about the forgeries by Attorney Hawthorne, who felt he had no choice in the matter: Bill would be arrested on a technical charge of murder and held in the courthouse jail pending an appearance before the Dade County Grand Jury.

Chubbie's first move was to go back over the events of the past few months and try to sort out some kind of rationale. If Bill hadn't done it, then Haden must have shot himself. With a lawyer friend, she went over the reasons her intended husband might have committed suicide:

Reasons why Haden Clarke may have killed himself

1. Remorse at the situation he had created, after his promise to Bill Lancaster.
2. Doubt of himself and of me. Fear that the past five years would prove too strong a bond and I would return to Bill.
3. Financial worries.
4. Doubt of his ability to write the book and make money with his writing. He talked constantly of this; his writings were all returned.
5. Intense sexual life over many years, suddenly discontinuing.
6. The fact that he was very young and I had placed too much burden and responsibility on him.
7. Physical condition.

8. The fact that he was very temperamental and emotional; that he rose to the heights of joy and sank to the depths of despair.

Were they believable? Chubbie thought so, but was wary of spreading the list around since she didn't want to hurt Haden's mother. It probably wouldn't have made much difference anyway; when Lancaster appeared before the grand jury he was indicted for first-degree murder. If found guilty, he faced electrocution.

Chubbie had a premonition that Bill would not do anything to defend himself; he would just sit in court, an upright and decent Englishman, and tell the jury he hadn't done it and if they didn't believe him, well, that was their problem. It was the one thing he knew how to do: behave like a gentleman, even if everyone else thought he'd acted like a killer. But come to think of it, from his point of view, didn't a bounder like Haden, drug-taking, diseased and dishonourable, deserve the favour of a quick bullet to the brain? There were probably a good many Dade County natives who, as they read the headlines, would have agreed that Haden Clarke was lower than a snake's belly and had got no more than he deserved. However, this was unlikely (not impossible in Miami but unlikely) to be the view of a jury and if they did return a guilty verdict, the judge would have no choice but to sentence him to the electric chair.

Chubbie went to a legal acquaintance, James 'Happy' Lathero, and asked if he knew of a really good defender who might take Bill's case. The only possible choice, Lathero said, was James M. Carson, a local attorney with a national reputation; a charismatic bear of a man who boasted that no client of his had ever spent a day in jail after he'd represented them at trial. The diminutive Chubbie went to see him. He was, she agreed, rather terrifying, looming over her, his face red and creased, as he told her: 'Lancaster is as guilty as hell. I wouldn't touch this with a bargepole.' She wasn't intimidated. She asked him simply to suspend judgement and go and see Bill in the county jail and make up his own mind. 'Why should I?' he asked. 'Because,' she said, 'He isn't a coward.' Something in that answer fascinated Big Jim Carson enough to make him cancel his appointments and go down to the courthouse to meet the prisoner. He *was* impressed. Not convinced yet; that took three days investigation of all the available facts in the case. Then he called Chubbie back to his office and asked her to sit down.

'We have got to get him off,' he said, 'But it's not going to be easy because of those damned notes. He shouldn't have done that. Also he is a foreigner, British, and even I don't know how it's going to go. The only way I can see is to make you the scarlet woman, the cause of it all. Can you take it?'

Chubbie didn't have to think about it. Bill was a friend still, if not a lover; Haden . . . she still hadn't sorted out what Haden had been, beyond a few moments of Miami madness under a big yellow moon; Bill was a friend, a mate, and she had never let a mate down and wasn't about to start now. 'Yes,' she told him. Carson then agreed to take the case. It was fairly obvious that neither Chubbie or Bill had any money, in all likelihood he wouldn't get paid but the publicity, if he could bring it off, would more than cover the bill.

Chubbie, as a material witness – she was being called by the prosecution – was not allowed to visit Bill, but he sent a note asking her to walk by the court at a certain time so he could see her from his cell window. He wrote that he was pacing his cell to keep fit and writing a long letter to Kiki, explaining the situation, and also to his elder daughter, assuring her that although they all said he'd shot a man and killed him, it was not true. As soon as Jim Carson heard about the walk-by he put a stop to it; if the press got hold of it they would make it look like collusion.

As the trial date drew near, Chubbie was forced to move out of Coral Gables, because too many people were coming by to look at the notorious woman in the big case. She also had to go through hours of questioning by the prosecution, covering every second of the afternoon and night of the death. When the trial started, as a witness, she was not allowed in the courtroom but had to wait until the end of the day before finding out how things had gone.

The trial opened on 2 August 1932. Crowds gathered outside the county courthouse trying to get a place in the sixth floor courtroom. The press had been headlining the trial as a legal spectacle; this was Miami noir but for real, and no one wanted to miss the excitement. After the jury was chosen, the lucky spectators were let in. They saw Bill Lancaster sitting at the defence table between Jim Carson and 'Happy' Lathero, all of them trying to look relaxed in the face of a barrage of camera flashes as the press got their pictures. The prosecution sat across the court at their table, represented by state attorney Hawthorne and his deputy, Henry M. Jones. Also at the table was Dr Beverly Clarke, Haden's brother.

Carson started proceedings by asking for an adjournment, on the grounds that his main medical witness, Dr P. L. Dodge, an expert on suicide cases, was ill and unable to attend. The request was denied by the judge. Hawthorne then opened for the prosecution.

He said that Bill Lancaster was obsessed with Chubbie, as his diaries would prove (Chubbie knew nothing of them) and was devastated when he received the news that the one man he trusted in Miami, Haden Clarke, had betrayed him with the one woman he loved. Stuck in Mexico with two business associates (Tancrel and Russell), his first thought was of getting home. He borrowed a plane and flew east as fast as he could, buying a pistol and a box of ammunition in St Louis. In Atlanta, the last stop before Miami, he opened the box and loaded the pistol. That evening, on arrival in Miami, he seemed to accept the marriage plans of Mrs Miller and Haden Clarke, even offering them a gift of $1000 on their wedding day. After dinner, however, he began to argue with Haden, causing the writer to jump up from the table and clench his fists in anger. At that point, Mrs Miller left the room. After more conversation, Lancaster and Clarke went to bed in the large sleep-out room at midnight. At 4 a.m., the alarm was raised. The police found no fingerprints on the pistol; it had been wiped clean. Two suicide notes 'alleged to have been written by Clarke' were later shown, and admitted to have been typed by Lancaster, who also asked a friend (Huston) to say that the pistol in the room had been his. The killing had been premeditated; there were clear attempts to hide the facts after the event. There could only be one conclusion: Bill Lancaster had flown back from Mexico to kill his love rival.

Jim Carson replied for the defence. He admitted from the start that the notes had been forged after Lancaster had woken up to find Haden Clarke dead. It was a stupid thing to do, but understandable, if the jury considered that Lancaster still hoped to convince Chubbie to come back to him. Stupid but not murderous. He went over Lancaster's record as a flyer and record-breaker and said he would call witnesses to testify to the man's honesty and decency. Since the defence would be calling Tancrel and Russell to give evidence of Lancaster's state of mind before flying east, he would be sure to ask the two of them for their temporary addresses. (They were both in custody on smuggling charges.) The defence would show the jury, Carson went on, that Lancaster had flown back hoping to sort out the situation and that, by the end of dinner that night, it had been agreed by all three that the wedding should be postponed a month. Yes, there had been tensions, Haden Clarke had become extremely bellicose, but by the time the two men retired to the sleep-out, the situation had calmed down.

At some point during the night, Lancaster was awakened by a loud noise. He jumped out of bed, turned on the light, and saw Haden Clarke lying with his head on a blood-soaked pillow. He raised the alarm after writing the suicide notes. Yes, he had lost his head for a moment but, no, he hadn't killed Clarke; who was, after all, a deeply worried young man. He was a heavy drinker, trying to cut it out to please Chubbie; he was a marijuana smoker; he had a venereal disease which was delaying the marriage; he was failing at his chosen career as a writer and he had discussed suicide with several friends in the preceding weeks.

Over the next few days the prosecution called their witnesses. Ernest Huston, the attorney, testified that he had lent Lancaster a pistol but that the gun used in the shooting was not it; he also agreed that he had been asked to pretend ownership of the weapon but had declined to do so. For the defence, Jim Carson then showed him a picture of Lancaster's bed taken by police when they arrived, and asked if that was how the bed had looked to him when he saw the room before the police. He said it hadn't, the pillow had been slept on, but in the photo it appeared untouched. Carson asked if Lancaster had said that he wished Haden could speak at least enough to tell them why he'd done it. Huston agreed that he had.

The ambulance team described how they had taken the still-living Clarke to hospital; the detectives detailed their investigation of the crime scene.

The prosecution then called Chubbie, Mrs Keith Miller as she was still known, to the stand. State Attorney Hawthorne paused while the court photographers shot a few hundred pictures. It did nothing to calm Chubbie's nerves, but then it wasn't intended to. She was asked if she had been engaged to Captain Lancaster.

She had not, since he was still married. She did, however, intend to marry him but then she'd met Haden Clarke. Yes, she thought Bill did suspect that things were not right between them and although she knew he would be upset by the news, she didn't think there would be trouble.

On the night of the shooting, what had happened?

She told the court, and added that she thought then and still thought that Haden Clarke had shot himself; he was very volatile, she had often seen him lose his temper with friends and his mother, with whom he had a particularly fraught relationship. However, on that night, even though she had left the room when there was tension between the two men, she didn't think Haden had been in a violent mood. He hadn't been drinking and that was what usually fuelled his worst excesses.

Hawthorne asked if she wanted to save Bill Lancaster. She admitted that she did. Would she lie to do so? No. She would sacrifice herself but she would never let a lie stand; why else would she have brought the forged suicide notes to the attention of the police?

Carson then had his chance to question the witness. He took Chubbie through her relationship with Haden, pointing out the writer's earlier marriage, his lies about it, his fury when he discovered his divorce was not final, his desire to marry at once despite his venereal disease. He asked about various documents, letters and cables which had been in the house and were now missing, because of, he was able to show, some extremely shoddy police work. It was clear, he told the jury, that the prosecution's case was undermined not only by the circumstantial nature of the evidence but also by hurried and careless police and medical investigation.

Carson next took Chubbie over the events leading up to the death of Haden Clarke. She said that before she'd gone to bed she and Haden had spent a few minutes alone together. 'We talked again of the awful situation and then I made a fool suggestion about suicide.'

What she had said was that Bill was so hurt and everything so awful that maybe the best thing would be for her and Haden just to kill themselves to make it all stop. Haden had told her that he felt like that too. Then she went up to bed and, as she went, he told her to make sure she locked her door.

What then?

'Bill was banging on the door, telling me to get up. I ran to Haden and asked him to speak to me. He was . . . moaning and groaning and his feet were striking against the iron end-rail of the bed. So Bill took the pillow from his own bed and placed it under Haden's feet to keep them from doing this. The ambulance man who had arrived said not to do it as it would make the blood rush to his head.'

Chubbie found herself shaking with nerves as she was questioned; she was terrified that in some way she'd put her foot in it. She knew that lies simply would not come out of her mouth convincingly; she was stuck with telling the truth. Her throat was dry, her voice often seemed to fade away.

The judge, sitting to her right, would lean forward at these times and ask her, quietly, to tell him her evidence and then, once she'd got it out, to repeat it more loudly to the jury. The one thing that kept her going was the sight of Bill Lancaster sitting at the defence table as if there wasn't a thing on his mind. He was absolutely calm.

Next was the medical evidence from Clarke's family doctor who had attended him at the hospital. This was pretty straightforward and unchallenged by Carson.

The boys from Latin-American Airways were then called. Russell talked about Lancaster's continuing anxiety over Chubbie's precarious financial position and his own inability to support her. This allowed the prosecution to introduce Lancaster's personal diaries, in which he had written at great length about his devotion to Chubbie and the pain caused by his inability to get a divorce and marry her. Readings from the diaries took a couple of days and showed just how obsessive Lancaster's relationship with Chubbie had become; they also showed, through everything, what a decent, even naive, but thoroughly honourable man he was.

Tancrel was called and retailed a conversation he'd had with Lancaster, during which the pilot had speculated on Haden Clarke's betrayal and said, 'I don't think he's double-crossed me, but if he has, I've seen a lot of dead men and one more won't make any difference.'

The embalmer was called and told the court that he had not noticed the powder burns on Clarke's head he would have expected after a suicide. Carson tried to counter this by drawing attention to the embalmer's lack of medical credentials. He did the same with the next witness, the detective who'd answered the emergency call, Earl Hudson. He had picked up the gun, which he testified had blood on the grip, wrapped it in a handerkerchief, and put it in his pocket. Hawthorne had attempted to show that Lancaster had wiped the gun clean of fingerprints after the shooting, Hudson's evidence suggested that it had been wiped by accident in the pocket of an extremely careless police officer; one, moreover, whose sister-in-law had been a girlfriend of Clarke's in the past.

The second week of the trial was to begin with Chubbie again on the stand, facing cross-examination by state attorney Hawthorne. She had been nervous before, when giving a straight account, but had come across as an honest witness, even if biased towards the defendant; she knew that Hawthorne would be out to show she was unreliable and use her evidence to put doubts in the minds of the jury.

Did she, in her first statement (given on the night), assert that Lancaster could not have written the suicide notes?

She did.

Did she state that Lancaster's sense of honour would not allow him to do so?

She did.

So the first she knew of the forgery was when Bill Lancaster told her?

It was.

If he had told her that he had killed Clarke, would that have surprised her more than the admission of forgery?

Certainly it would.

Even though she had stated, at the time of the killing, that she was sure both that Lancaster had not forged the notes and had not killed Haden?

There was no answer to this. After a pause to let it sink in, Hawthorne asked if she still loved Haden Clarke.

She didn't.

What about Bill Lancaster?

No.

When did she stop loving him?

About two years ago. But I am still very fond of him.

Is there any reason why love died? Hawthorne asked.

No.

Then you quite deliberately let him believe you still loved him, you said nothing at a time when he was doing everything he could to beg, borrow and steal a dollar or two to support YOU?

Chubbie once again said nothing. She knew quite well that for years she had been keeping Bill. Hawthorne repeated the question and added:

'Weren't you a deliberate traitor to Lancaster in all those letters, all those times you said, "all my love to you"?'

He didn't understand their relationship, Chubbie tried to explain. They were friends, more than friends, they had been through so much together that . . .

And Hawthorne left it to the jury to complete the sentiment, that she would have done anything to save Lancaster, so surely her evidence, her 'belief' in his innocence was worth as much as her avowals of 'all my love'.

The prosecution turned its attention to Chubbie's relationship with Haden. She maintained that his constant lying had destroyed any feelings she'd ever had for him. Hawthorne hammered away at this for a while, then returned to Bill Lancaster, stating that one of the things Chubbie had said she admired about him was his code of honour. She agreed. Then, trying to tar Lancaster with the same pitch that had been applied to the dishonourable Haden, Hawthorne asked: 'He'd steal for you, would he?'

Chubbie denied it.

'Didn't he steal a chicken for you?'

Chubbie paused moment, then said, 'It was a duck.' The courtroom erupted in laughter. The answer somehow defused the tension into humour, and made Bill look funny rather than dishonourable, and Chubbie even more honest. But Hawthorne wasn't finished. He turned off the mirth by snapping out, 'Do you know Lancaster has a wife?'

Chubbie managed to get out no more than: 'Yes, but . . .'

'From whom he is not divorced?'

She tried again, but Hawthorne was having none of it.

'And two little girls?'

It made Lancaster look like a deserter and a rotten scoundrel. Hawthorne asked if Chubbie remembered an occasion when Lancaster had pretended to be driving a car she'd been drunk in charge of; was that lie a decent thing? At this point, Jim Carson objected, and the line of questioning was stopped. However, damage had been inflicted and the prosecution built on its success by digging further into the confused relationship between Chubbie and Bill.

'You said earlier that you intended to marry Lancaster?'

'Yes, I always felt that when Bill was free from his wife in England I would marry him.'

'But you weren't in love with him?'

'Being in love and just loving a person are two different things. I was not thrilled or infatuated with Lancaster, just terribly fond of him.'

'Were you infatuated with Clarke?'

'Yes.'

'Now you do not even love his memory?'

It was a brutal question to ask someone who had been through, and was going through, as much as Chubbie. She murmured, 'No. Unfortunately no', and gave way to tears.

Jim Carson's cross-examination was brief, merely underlining Haden Clarke's unstable temperament, and bringing out the fact that Chubbie had been and was still paying out £30 a month to Kiki Lancaster and Bill's two daughters.

Technical witnesses were called or recalled and while Hawthorne laid out what he felt was damning physical evidence, Carson tried to muddy the waters. By that Monday evening, the State had finished its presentation. Bill Lancaster would take the oath first thing in the morning.

Released from the ordeal of cross-examination, Chubbie was unable to relax. Denied access to the court, she had to wait outside while Bill fought, quite literally, for his life. The only thing she could be sure of was that Bill would be Bill and tell events exactly as he saw them, whether it served his case or not.

The first day was filled with his life story from war-time flyer, through his marriage and its breakdown, his meeting with Chubbie and the Australia flight and their later adventures in America. Character witnesses were called who testified to his honesty and courage. The second day was filled with much of the same, as if Carson wanted to bludgeon the jury into acquiescence with the view of Lancaster as the very model of an English gentleman. He did, however, ask him, baldly, 'Did you kill Haden Clarke?'

'No, I did not.'

Lancaster went though the history of his acquaintance and then friendship with Clarke, which allowed Carson to bring out a number of unsavoury facts about the writer, since the two men had shared the sleep-out whenever Bill was in residence. He had seen Haden with at least three different women overnight; he hadn't ever seen him actually working (though in a very Bill move he blamed himself for that, since Haden always wanted to come walking with him); he had seen him in his rages and his cups and giggling hysterically after smoking marijuana. He recalled the last conversation the two had had before Bill set off for Mexico, when he had asked the writer to look after Chubbie. Haden had said, 'I will care for her in such a way as to make you remember my friendship for ever.'

Carson moved on to Kiki Lancaster. The prosecution objected but Carson pointed out that they had introduced the subject; and the judge allowed him to continue. He was able to elicit from a reluctant Bill that Kiki was Roman Catholic and would not countenance a divorce under any circumstances. This wasn't quite true. According to Chubbie, she had offered to do it for £10 000, but it did show Lancaster in a less unflattering light, as victim of the situation rather than deserter.

The testimony of the Latin-American Airline boys was branded totally false and their criminal records did little to counteract that view.

Then came the moment the press and spectators and jury were waiting for: the night in question. Lancaster had arrived home with a loaded gun. Why?

Because Ernest Huston had lent him a loaded gun, he'd lost it, and he wanted to return the replacement in the same state. Besides, what use was an unloaded gun?

Did he talk alone with Chubbie on that night?

No. She had refused.

After Chubbie had gone to bed, what did Bill and Haden talk about?

Lancaster appeared hesitant. He said that he'd purposely not said anything about the conversation to anyone so as not to upset Haden's mother. Carson reminded him, pretty sharply, of the situation he was in and the necessity for telling the whole truth. If it was a set-up, it was a brilliant one, allowing Bill to gain the moral high ground before delivering the low-down on Haden Clarke.

They had talked about Haden's venereal disease. The writer had shed tears of deepest regret over the situation that he and Chubbie were in. He had gone on to say that he'd had many women, but this was the real thing at last. He was truly in love with Chubbie and would do everything in his power to make her happy. He was, however, worried about his ability to succeed as her biographer. The two of them went to bed, with the last words coming from Haden, who said, 'Bill, you're the whitest man I ever met.'

The last time Lancaster had seen the gun, it had been lying on a table between the beds. He went to sleep and was woken by a shot. Haden was lying, moaning, with blood on his lower jaw. Lancaster asked what had happened. It flashed through his mind that Chubbie might think he'd done it out of jealousy, so he wrote the notes, then took them over to the bed with a pencil and asked Haden to sign them. There was no response beyond groans. After that he went and called Chubbie.

It was the turn of the prosecution to cross-examine. After asking Lancaster who had killed Haden Clarke and getting the answer, it was suicide, Hawthorne concentrated on the notes. He held the first up:

'This note written on Latin-American Airlines stationery. Did you write it?'

'Yes.'

'Positive?'

'Yes.'

'And this note addressed to Chubbie. Is that your work?'

'Yes.'

'Positive?'

'Yes.'

'Are you as positive as you were on April the second that it *wasn't* yours?'

Hawthorne went through the buying of the gun and the ammunition and the loading of that ammunition in Atlanta; he wondered why, if Lancaster was so concerned about Chubbie being left without funds – and the words were in his diary – did he spend more than $30 on a pistol and ammunition? There was no answer to that. However, when Hawthorne asked what Lancaster thought of the relationship between Chubbie and Haden, he surprised everyone by answering that in many ways it was quite beautiful.

Even after he'd found out about Clarke's venereal disease?

'Even after. They were very much in love, there was a better side to Haden's character and I saw only that.'

'He proved that by committing suicide, didn't he?' Hawthorne asked.

Bill said, 'Not in the manner in which he did, but the fact that he did showed he had good intentions.'

'Do you still feel that way?'

'I would like to.'

'Did you feel relieved when you dropped off to sleep that night?'

'Yes, in my heart I knew that Clarke would never marry Chubbie.'

'Why? Because he would commit suicide before morning?'

'No. Because he had promised to tell Chubbie in the morning all of the mis-statements he had admitted to me that night, and if he didn't, I would.'

After spending some time on Tancrel and Russell's evidence, trying to revive their reputations, Hawthorne went back to the night in question and the faked notes. It wasn't a particularly productive line of questioning, since Lancaster had admitted to faking them, and the admission tended to offset the initial dishonesty. All the way through the cross-examination, Hawthorne's tactic was to cast doubt on Lancaster's so-called honour, knowing that if he could destroy the aura of decency that clung to the man, the verdict would be certain. However, he couldn't manage it, even when he quoted a diary entry in which Bill criticised American justice as being 'wet', the flyer was able to regain the court's sympathy by his open apology for the sentiment and his concern that it might have caused offence. He was allowed to leave the witness box after 12 hours.

Jim Carson now called a series of witnesses to show that Haden Clarke was a bad lot; a thief who stole money from his friends; who caught venereal disease from who knows where and didn't give a damn if he passed it on (and had said as much in front of a witness); who had made a habit of talking about suicide as a convenient way of getting out of difficult situations and had studied it enough to believe (wrongly) that a shot to the head was the best method. He had deserted his wife and had affairs with a number of other women, to none of whom had he been honest or faithful.

It was not difficult to blacken Clarke's character. He was a confused, mixed-up young man with too much charm for his own good; he'd never had to struggle for anything and, the moment he did, for his writing or his love for Chubbie, he seemed to collapse; he was a natural follower and young for his age. Suicide is a young man's vice, but he had walked away from relationships in the past with hardly a thought of the pain he might cause and, surely, it would have cost virtually nothing to his self-esteem to walk away from Chubbie. Bill, on the other hand, was a different matter. From where Haden stood, Bill was the real thing. A pilot and adventurer, a real 'white' man who would certainly kill himself before committing a dishonourable act. What did the fatherless Haden really want: Chubbie's love or Bill Lancaster's approbation and how far would he go to gain either?

In rebuttal, Hawthorne called a number of Clarke's friends who denied he was given to rages or depressions and stated that on at least two occasions he had told them there would be trouble when Lancaster returned to Miami and that he was anxious about the future. The prosecutor also quoted a supposed remark that Lancaster had made, soon after Clarke's death, to Clarke's mother: 'Sometimes I think I did kill Haden.' The State was going to call Mrs Clarke to the stand to elaborate on this conversation but never did, sighting her emotional condition.

The medical evidence looked in detail at the shooting. There had been, through some astounding ineptness, no autopsy at the time of the death. However, the body was exhumed and the skull removed and exhibited in court. Defence experts were called to show that, contrary to the State's evidence, there *were* powder burns present in the bone and that this indicated the gun muzzle was held against the head, which would be consistent with suicide. The bullet path was clear, entering at a point midway between, and three inches above, a line drawn between the right eye and ear, and exiting below the left ear.

The prosecution maintained that the wounds were perfectly consistent with a second person holding the gun muzzle tightly against the head and firing. There was no proof either way, as throughout the trial; it all depended on who you believed.

After two weeks it was time for final statements. Jim Carson went through the State's case witness by witness, casting doubt and uncertainty, emphasising time and time again that this was a case about a man's life, not about a speeding fine or a lie; it was a case about character and the lack of it; in the end it was all about who you believed: Captain (actually, still Flying Officer) Bill Lancaster, who had sat quietly, day after day, obviously certain of his own innocence; or the memory of Haden Clarke, living off women, a depraved drug fiend, spending his time talking about poetry or collecting unmentionable social diseases and losing his temper in a most unmanly manner. The physical evidence, as the defence had shown time after time, could point in either direction or neither because there was not one shred of it that could say, concretely, here is a murderer.

State attorney Hawthorne rubbished Carson's summing-up. The State's witnesses had, he said, made a highly convincing case. Not all the witnesses were nice people; you wouldn't necessarily want to ride a streetcar with them, but that didn't alter their evidence and the jury should remember that they were all associates of Lancaster. Men and women he'd chosen as friends or colleagues. What did that say about his judgement? Then there was Lancaster's diary. It spoke of a great love and no one who had heard it read, as they all had, could fail to be moved by those words; and neither could they ignore the threats written down in black and white. No one was saying that Captain Lancaster was a coward: quite the opposite, he was an adventurer, the kind of person who scorned convention and went his own way, beholden to no man or law, particularly the laws of a nation he'd characterised as 'wet'. Was he really a character who would tamely sit back while someone like Haden Clarke lorded it over him in his own home and stole the woman he loved? It was scarcely credible.

The forensic evidence pointed one way only. No one would lie in bed and hold a gun at an unnatural angle against the top of their head to kill themselves. Simply exerting enough pressure on the trigger at that angle was virtually impossible. It was far more likely that Lancaster, who had seen many dead men in his time, simply stood over his sleeping rival and executed him, as he would a sick dog. The case was not about sympathy or emotion, it was about one question: did Haden Clarke kill himself or was he murdered by William Newton Lancaster?

The jury spent four and half hours over their decision, coming back once to look at the exhibits in evidence again and to ask the judge what exactly constituted reasonable doubt. Still waiting outside, Chubbie could see the crowds of reporters and spectators milling around the courtroom door. Then they were called back in and the doors shut. She waited and waited, and finally there was a huge cheer and she knew it meant acquittal. She didn't go in; that would be wrong, she thought. She made her way downstairs and out the back of the courthouse. Only one reporter, arriving late, recognised her. He asked for her reaction. She said she was delighted, she knew old Bill would come through.

She didn't go back to her hotel but went to 'Happy' Lathero's house where she found Bill, Jim Carson and Lathero had already arrived. The relief of that moment was indescribable. Carson told her that the judge had said the verdict was the right one. He said he'd never had any doubt that they'd do it, but he was a lawyer and Chubbie could never be quite sure just how confident he'd been. Maybe it was the sight of Bill sitting there all those long days that had given him the confidence he'd needed. As for Bill himself, the sole earthly witness to the events of that night, he seemed unchanged; thinner, pale from the months spent in the courthouse prison but, essentially, exactly the same man he'd been when he was first arrested.

Bill Lancaster went straight back to England; his notoriety meant there would be no job for him in America. Chubbie closed up the Miami house and considered her options. She might find some work flying but if she did it would only be as a curiosity. The Scarlet Bird perhaps; not something that appealed to her. She was young still; there were many possibilities for her but, like Bill, she wanted to get out of America. There were too many memories to haunt her there. On the other hand, Bill was in England, and so were his wife and children and his parents. It was his country, not hers; but then, did she want to see him?

After the trial he'd been asked if his future plans included Mrs Keith Miller; he'd dodged the question. What if she'd been asked the same question? She knew what the answer would be: no, they didn't. Not in the long term. Bill was obsessed with her; he could never be with her just as a friend, he'd always want more, even if he tried to hold back; sooner rather than later he'd plant a briar hedge around his princess and water it and tend it until she was protected from all the world except him. Only she'd never wanted to be protected from anything or looked after or treated like . . . something precious. She'd almost given way to the urge once before with Haden, and look where that had led them all. And then there were the little white lies she'd caught Bill uttering over the years; had they grown too; was he capable of a big, black lie? And she'd never really known exactly what his relationship with Kiki had been. He'd once told her that Kiki had been involved with another man when Bill had been posted to Austria after the war; he maintained that his youngest daughter was not, in fact, his child. Was it true or was it another white lie to make her feel better about their adulterous relationship?

She went back to England by ocean liner at the end of 1932; it was a very different arrival from her first as a young Australian girl; now she had to be sneaked off the ship in London's George V docks to avoid the press. She was met by Bill at the quayside. It was a subdued meeting. He was living with his parents in Sydenham. Chubbie took a bedsitter in Edgware and started to look for work. She made some contact with the *Daily Herald* who suggested that a reporter ghost her autobiography for the paper, which at least gave her a financial breathing space while she sorted out her future.

She and Bill would meet every so often to go to clubs or out for meals. They tended to be discreet, although they couldn't always avoid trouble. One winter night when Bill was escorting her to the car, a shrouded figure rushed up and slapped her viciously hard on the cheek. Bill gave chase and returned some time later, saying it had been Kiki.

She had nothing to be jealous about. Bill and Chubbie were no more than friends, although Chubbie was gradually severing ties and, when Bill told her the only way he could see to earn a living during the Depression was by going back to record-breaking flights, she made it clear that he would be doing so on his own.

His new plan was to beat Amy Johnson's record flight time from London to the Cape. Lady Bailey had recently attempted to cap the record but failed; if Bill could succeed, it would get his name back into the headlines in a good way; he would be a contender again and, besides, it was better than working. He got his father to finance a plane, an Avro Avian. Presumably the chance of getting his son away from the claws of the scarlet woman had prompted Lancaster senior to help out, something he'd been reluctant to do in the past;

although there must have been times when he wondered if his wayward son would *ever* grow up.

Chubbie was concerned, as she followed the preparations from a distance. She knew how slapdash Bill could be; there had been many occasions back in the days they were flying together when she'd had to shout at him to get him to check vital mechanical and navigational points. There was also what she considered his crazy idea of flying the Oran–Reggan Desert leg at night. It would be cooler, but he had no experience of night-flying and the Avian had no night instruments; if he missed his airfield, he would find himself flying over one of the most inhospitable areas of the Sahara. Still, this was his flight, not hers. She wasn't even at Lympne Airfield in Kent to see him off. They'd met for a brief farewell in London two days earlier. He hadn't told her he'd made a will naming her as his sole beneficiary, not that he had anything to leave apart from an insurance policy on his life. He did, however, ask if he could wear her watch on the flight; she told him it would be useless, it was so small; he settled in the end for two photographs of her.

He took off before dawn and turned south across the Channel. The water was choppy down in the darkness and there was a nasty headwind blowing. He made first landfall at Le Havre, refuelled and took off again, hoping to make Oran without landing; the wind, which persisted, was slowing him down; he was already behind Amy Johnson's time and things were getting worse rather than better. He had to stop over at Barcelona and when he finally landed at Oran he was already four hours behind schedule. Refuelling and maintenance took up another two hours. Leaving Oran, he made for Adra, closer than his original destination of Reggan; he could refuel there and fly on, missing the Reggan stop, going directly to Gao and, he hoped, making up the lost time. What he couldn't make up was missed sleep. He was getting increasingly ragged and, on taking off from Adra, he lost the desert road he was following and wasted yet more time and fuel trying to get back on course. This meant he would have to change his plans again and land at Reggan anyway, for rest and fuel before making the longest of the desert hops.

It must have seemed as if the Sahara itself was against him, or perhaps warning him: shortly after arriving at Reggan, a sandstorm blew up. The local French depot manager told him he could not possibly take off in such conditions. Bill saw his record flight slipping away. If he didn't leave soon, there would be no point in leaving at all. The attempt would be over – the career he hoped to build on it would be finished before it started; there would be nothing left.

The storm began to abate just a little and he clambered wearily back into the cockpit. The manager told him he was crazy; the swirling sand would have wiped out the desert track, Lancaster would have to fly over 500 miles of featureless terrain by instrument alone and, in the dark, his only means of checking his course would be by lighting a match to read the compass heading. The manager begged him to wait at least until moonrise; Bill sat fuming with impatience at the delay but at last the big desert moon began to climb over the dunes and he nodded to the mechanic to spin the prop. The engine started with a roar and the Avian began to move forward, faster and faster until the wheels left the hardened sand and the ship rose up into the night sky and flew away into the moonlight.

Twenty-nine years later, Mrs Chubbie Pugh came downstairs to make breakfast in the kitchen of her house in Berkshire, England. As she was heating some milk, the phone began to ring. She called to her airline pilot husband Johnny to answer it. He did so, taking rather a long time. When he came through he asked her what she'd been up to lately. She said, 'Nothing.' He said the caller had told him, 'Chubbie's in the news today.' Neither of them had looked at the paper yet. When they did, they read a report that lost aviator Bill Lancaster had been found in the Sahara Desert by a French army patrol. His body was lying in the meagre shade of a wrecked plane; due to the dry conditions, it had become mummified. The flyer's possessions were lying beside the body; they had also been perfectly preserved. There was a passport, a wallet with two pictures of Chubbie in it, and a logbook in which the flyer had recorded his last days on earth without a word of self-pity. There were messages left to his father and his mother, and to the woman he had loved so.

7
Degrees of difficulty

*A*fter the Great War things began to change but by no means as fast as popular historians believed. The 1920s roared about as much as the 1960s swung – not a lot. Most of the roaring, like the swinging, was done by middle-class children who, when faced with a choice between ideals and income, voted for a comfortable future. But on the way there *was* a fallout and some things *had* changed. In the early 1920s, after that colossal struggle in Europe had finally staggered to an end, the ground gained allowed space enough for a few, very determined young women to climb into a plane and take off – sometimes because they believed they had a right to feel the joy of achievement, sometimes for the fun of it, sometimes, as Bessie Coleman said, to create the minimum of their desires.

Not that it was ever going to be easy for someone who had to fight a whole raft of prejudices – against women in the air; against black Americans anywhere near an aircraft, let alone flying it; against the black race itself, at least the male half of it, which, in the person of her brother John (who had served in a black regiment in France), was endlessly lauding the superiority of Frenchwomen and laid it down plainly: 'You nigger women ain't never gonna learn to fly like they do in France.'

Bessie realised that one half of this remark was true: she wasn't going to learn to fly in America. As for the second part, she knew it was up to her, if she believed in herself enough, to get to France, find a school and persuade someone to teach her how to fly. It wasn't an impossible dream: working in Chicago's South Side as a beautician, she was living in a culture where black businesswomen like Madame C. J. Walker and newspaper editors like the *Chicago Defender*'s Robert Abbott wielded both financial and political power. Of course, she also lived in a society where race riots could erupt at any time, as they did in 1919 when a young black swimmer strayed into a white-only area of Lake Michigan and rioters with flaming torches were soon out on the streets howling for blacks to 'come out and get your asses whipped or stay in and be barbecued.'

Bessie was born in 1892, in Atlanta, Texas, which was little more than a settlement founded by migrating Georgians from the city of Atlanta. Her mother Susan was black, her father George, half-Cherokee, half-black, both of them contributing to the remarkable looks which Bessie, in later life, was never loath to use to aid her progress in an inimical world. It was a time of change and opportunity, if you happened to be white; if you didn't, then Atlanta could be an uncomfortable place to live for a black family with ambitions of improving themselves. To be honest, anywhere in the American South was an uncomfortable place for blacks, with or without ambition; the journalist Ida Welles recorded 782 lynchings in the first decade of the twentieth century.

Shortly after Bessie's birth, her father bought land near the larger town of Waxahachie and moved the family to a house he'd built there. It was small, three rooms, but it had a yard and a porch and Bessie's upbringing was stable and happy among 12 brothers and sisters – she was number ten. During the cotton harvest, everyone in the town, adult and child, worked in the fields but for the rest of the year there was school and Bessie, who started attending at the age of six, made an immediate impression on her teachers. She was clever and she was hard-working. In the 'separate but equal' school (that is, separate but certainly not equal to white schools in staff numbers or equipment) she attended, she had little chance to extend herself beyond the basics but she did get an appetite for knowledge and a love of learning that was going to stand her in good stead later.

In 1901 George Coleman decided on yet another move, this time out of Texas where he suffered the double disadvantage of being black *and* Cherokee. His wife Susan decided not to follow him. This was at a time when almost 90 per cent of black families had a father at their head and the stigma of being a single mother – not to mention the financial difficulty of bringing up six children (the others had left home) – was going to be a real disadvantage. Susan Coleman was clearly a woman with a strong sense of purpose and no lack of courage; whatever the final cause of the split between the two, she was adamant that she was staying put. And stay she did, getting a job with a local white family as housekeeper and bringing up her daughters with a firm hand, reading to them every night, passing on tips about behaviour she'd picked up at her employers' and encouraging each girl to 'amount to something'.

With Bessie, this became something of a mantra – it wasn't long before she took over her mother's reading chores and entertained the rest of the family with what were virtually mini-dramatisations of the novels she read and commented on, offering: 'I'll never be a Topsy or an Uncle Tom!' on finishing Harriet Beecher-Stowe's novel *Uncle Tom's Cabin*.

After completing her education at the local school, she took her savings and enrolled in the Colored Agricultural University of Langton – but she could only afford to attend for one term and had to return to Waxahachie, where she took on a job as a laundress, saving money for what she hoped would be a successful escape from the little town.

Shortly after her 23rd birthday she headed north for Chicago where she got herself a room on the South Side, lodging with her brothers Walter, a Pullman Porter, and John, who hadn't managed to settle down after the Great War. Known as The Stroll, the area was, in all but name, a black ghetto, but it was also a place full of life and energy, where African-Americans could, as long as they didn't cross the boundaries and threaten white society, create a successful life for themselves. Bessie took a course at a beauty college and became a manicurist. She worked right in the centre of The Stroll and soon got to know her way around, meeting entertainers, politicians and businessmen and women as she attended to their nails with a skill and speed that was recognised in the columns of the *Chicago Defender*. She also got married in 1916, when she was 23; a marriage which is something of a mystery. The groom, Claude Glenn, was a quiet, elderly man with no business interests, no money, no obvious attractions at all – the couple hardly lived together, staying in the same house for no more than a month or two. Coleman biographer Doris Rich speculates that Bessie was looking for a friend; it's possible that she was also looking, rather as Pancho Barnes and Mary Heath did with their husbands, for a respectable married background which would allow her to exercise options that an unmarried woman would find it difficult to achieve in the society of the time without being labelled fast and loose. Glenn wouldn't make a fuss but would make a marriage, leaving her to make relationships with a number of the most powerful figures on The Stroll. Rumour speaks of a number of gentleman callers, both black and white, to her private rooms, and Bessie herself talked of a Spanish gentleman who was instrumental in her career in aviation.

Why flying, we don't know. Possibly because of her brother John's stories about France, the attitude there towards blacks, the excitement of aviation and the challenge he laid down, telling her she'd never be able to learn. She had always been a woman to accept a challenge, rather than look the other way. Perhaps she calculated that if she wanted to amount to something that would reflect both the aspirations of her own people and gain attention in the wider world of the United States and Europe – if she wanted to *become someone* – then the worlds of entertainment and sport were the only ones open to her. She wasn't a natural musician, writer or composer, she had no particular sporting talents; she did have, however, a brilliant and questing mind and a particularly fine grasp of mathematics; and if navigating and flying a plane was a matter of learning certain skills, she knew she could do it. But not in America.

She had tried enrolling in a number of American flying schools and even approaching individual pilots, but none would take her on either because she was black or because she was a woman or because she was both. In the end, she turned to a local acquaintance, Robert Abbot, the powerful publisher and editor of the *Chicago Defender*, the most important black newspaper in the country, asking his advice about learning to fly. A constant and vocal supporter of black rights in all fields, Abbot recommended she travel to France and enrol in a school; he told her, if she could learn French and save enough money for the journey, he'd make enquiries about teachers and terms and set up a course of lessons. Bessie didn't need telling twice; she enrolled in a French class and switched jobs, becoming the manageress of a highly successful chilli restaurant.

In 1920 she sailed to France and enrolled in the Caudron Brothers school for a seven month course. She later said it was a ten month course but, like a lot of flyers, there were many areas in which she found it was not necessary to be all that exact. Her age was another – like Katherine Stinson, she 'lost' a few years once she returned to the States and began to fly professionally.

One place where inattention was not permissible was the aeroplane itself: before every flight she scrupulously checked each bolt and spar and wire before taking off in the Nieuport '82 trainer on which she was learning. Flying itself, though chosen at random, became her passion; perhaps underlined by the greater freedoms of a country in which her colour was of no importance whatever (except to the American tourists, who objected to sitting outside cafés in her company). She said that she hadn't really lived until she'd experienced being alone in a plane thousands of feet above the earth.

She was awarded her licence by the Fédération Aéronautique Internationale in June 1921; after the ceremony she went on to take advanced lessons from a French war ace (though there is no actual record of this) then, after buying a complete wardrobe in the shops of Paris, headed for home.

The buying spree was prescient: she was going to need all those new costumes because she was a star from the moment she stepped ashore in New York. Newspapers of every hue were waiting to photograph the lovely young pilot as she walked down the gangplank; every black newspaper in the country had her picture on its front page; the *Chicago Defender* printed a long interview with the aviatrix, during the course of which she maintained that if blacks were going to keep up with the times, it was vital that young men should take up the challenge of flight. They would need fitness, courage and ambition and, since she could see these qualities all around her in black society, she could think of no reason why the dream of flying should not be achieved.

Thinking of her own career and looking at American women pilots, Bessie realised that if she were going to deliver on all the publicity and use it to earn a living, she'd need the skills to perform spectacular stunts at air shows – skills she didn't yet have, despite those lessons with the war ace. She would need more tuition and more time in the air; in short, she needed to go back to France. Which she did, this time, it is generally assumed, with the direct support of Robert Abbot.

Back in Paris, she found the atmosphere somewhat less congenial, with American tourists and expatriates exerting an increasingly baleful racist influence. It was a small nuisance but nonetheless served to remind Bessie that whatever she achieved, it was never for her own satisfaction alone but was always part of a larger struggle. Sometimes she must have felt the desire just to be herself, to achieve her own ambitions – it was never going to happen.

After a course of advanced lessons she toured Europe, meeting Anthony Fokker, the Dutch aircraft builder, and a number of German aces, including Ernst Udet, who had flown with the Red Baron, Manfred von Richthofen. The time away was to prove useful once she returned to New York, since she could create whatever legends she needed to underline her career, using experiences and quoting characters who were unlikely to contradict her words. And she realised quite clearly that to make a living, she would need to be larger than life – larger even than being the first black woman aviator, and only the second Afro-American to gain a licence (Eugene Bullard, an American who flew for France during the Great War, was the first).

So she set about becoming nationally famous. She appeared at a New York show organised by the *Chicago*

Defender, ostensibly to honour a black army regiment but, in fact, to publicise herself, which it did satisfactorily, though she didn't fly her own plane. A few weeks later she flew a borrowed Curtiss Jenny at a show in Memphis, drawing crowds of up to 20 000 over three days. Reporters commented on her skills and, though she performed no stunts, billed her as the Greatest Woman Flyer in the world, and to the all-black audiences who watched her, she was certainly just that. Those in the know, including the Curtiss Company pilot who delivered the ship, commented that, though conservative, she flew with considerable skill. The press, up to its old habits, immediately dubbed her Queen Bess, Daredevil Aviatrix and, she didn't mind one bit.

However, the press is fickle and for the black newspapers of the 1920s there was a new issue claiming their attention: just as in white male society, black men were not happy at the way their womenfolk were wandering away from the cooker and the cradle; female schoolteachers and professors, businesswomen and health professionals were all castigated for not having enough, or even any, children. It could lead, as one somewhat shrill editorial had it, to race suicide. And there's no doubt that Bessie had to fight this attitude herself, from the very people who, she felt, ought to be supporting her struggle. She wanted to establish a flying school to get black pilots into the air, but gaining sponsorship and raising money was no easy matter. She was able to buy one plane, an old Curtiss, from the army but though businessmen were happy to see her and newspapers only too glad to interview the 'neat-appearing young woman', promises tended to evaporate when it came time for them to be honoured.

Her ambitions weren't helped by a crash in California while she was flying before an audience of 10 000; her only plane was a write-off and she broke a leg and a number of ribs and was kept in hospital for some weeks while her broken bones set. Once she was out, Bessie found herself without a plane and without any more support from Robert Abbot at the *Chicago Defender*. In the first hectic months of her career, he'd provided her with a desk in the paper's New York office and employed a manager to set up future engagements; she, strong-willed and seeing no reason why she shouldn't have a voice in her own affairs, employed managers of her choice and set up a number of schemes on her own behalf, including a film about flying in which she would feature. Unfortunately, when her ideas didn't gell with those of the production company, she wasn't in any mood to back down; the project, and half a dozen others, fell apart, leaving her without money or management.

Then in 1925 she went back to Texas for the Juneteenth celebrations of black emancipation (Texas had been a year behind everyone else in honouring Lincoln's Emancipation Proclamation). She was to fly a borrowed Curtiss Jenny at the Houston air show and, as a Texas woman, attracted a huge crowd from all levels of society. And she flew like a dream, like an angel, like Queen Bess, thrilling the crowd to such an extent that even the white press forgot itself enough to praise her performance to the skies – though not enough to print the word 'negro' with a capital 'n'. Bessie also, shrewdly, offered free flights to local black ministers, who could provide a publicity network superior even to that of the *Chicago Defender*.

Texas skies were to prove profitable and for the rest of 1925 Bessie was in demand at shows and exhibitions across the state; she even tried parachuting on a couple of occasions and gave an exhibition over her childhood home at Waxahachie.

For the next few years her career went well; she designed a special uniform for her shows and managed to find both sponsors and supporters and though there were times when the work ran out and she found herself back in the beauty business, on the whole she felt pretty pleased with the way she was establishing her name as an aviatrix. The establishing of a flying school, which had always formed a major part of her ambitions, seemed as far away as ever but her example to young black men and women was laying down a foundation that would, in a few years time, lead to real advances in black aviation.

In early 1930, Bessie was contacted by a young black businessman and civil rights worker, John Betsch. He was in charge of the Jacksonville Negro Welfare League annual clebrations and, as an aviation enthusiast, he could think of no better attraction than Bessie Coleman. She agreed to fly and parachute jump from the Curtiss Jenny she'd just bought – it was being flown up from Dallas by William D. Wills, a delivery pilot, and would be in Jacksonville in time for the air show. The plane was far from new and a long way from being in the best condition; Wills was a fine pilot and it was all he could do to keep the ship in the sky for the journey between Dallas and Jacksonville.

Recalling her training days in France, Bessie wanted to give the Curtiss a test flight before performing in it and jumping from it; early on the morning of Friday 30 April, she drove out to the airfield with Betsch and met up with William Wills beside the ship. She looked it over, said a quick prayer and then she and Wills climbed aboard, the delivery pilot taking the front cockpit from which he would fly the plane while Bessie looked out from the rear cockpit to find a good dropping zone for her parachute descent. The Curtiss took off and climbed to 3500 feet where Wills began to circle the airfield and the immediate area; Bessie was leaning over the cockpit edge observing, when the plane suddenly went into a tailspin, plunging down towards the ground. At about 500 feet, the ship turned upside down and Bessie was thrown out of the cockpit and fell to earth. Wills, who was strapped in, tried to regain control but was unable to steady the craft and after hitting some trees it smashed into the airfield. John Betsch was among those who ran to help – it was obvious that Bessie could not have survived her fall. When he got to the plane, he was so shocked that, without thinking, he lit a cigarette – the leaking fuel caught fire and the wreckage exploded. It was not known whether Wills had survived the crash or not – he didn't survive the inferno.

Bessie Coleman was buried in Chicago. More than 10 000 mourners filed past her coffin as it stood overnight in the

funeral parlour before being moved in the morning to the
Pilgrim Baptist Church.

If flying was difficult for Bessie Coleman because of her race
and sex, it was not that much easier for Hanna Reitsch
because of her nationality. During the 1920s and early
1930s, while Germany was climbing out of inflation and
revolution into fascism and total destruction, young men
and women who wanted to fly were more or less restricted,
by the terms of the Versailles Treaty, to exercising their
talents through unpowered flight. Gliding had become a
popular sport; it was also a training ground for future
fighter and bomber pilots and although Hanna never
actually went to war, she was to fly as close to it as any
woman had so far. It was to be quite a journey: as a girl she
wanted to become a flying doctor but settled for being a
highly decorated Nazi test pilot, the last person to fly in and
out of Berlin as it crumbled under the shells of the Russian
army.

Born in 1912, the daughter of an eye surgeon, there never
was a time when she didn't want to fly. Birds, clouds, even
insects seemed to be forever beckoning her towards a future
with her feet firmly off the ground. Her family was of a
different opinion: feet firmly *on* the ground treading the path
of marriage, lots of children, lots of respectability was their
plan for young Hanna; she had different ideas. She also had
common sense, and planned a campaign that opened, when
she was 12 or 13, with the dawning desire to become a
doctor. Dr Willi Reitsch, who had often delighted in bringing
home an animal eye from the market to entertain the
children with an off-the-cuff dissection, couldn't really
object. In fact, he was proud that Hanna wanted to follow
him into medicine. Then she sprang the trap: she wanted to
be a flying doctor. Willi was not pleased, so Hanna added
missionary to flying doctor. By now, Willi Reitsch must have
felt he was facing a *fait accompli* and conducted a fighting
retreat: if Hanna could complete her school studies and do
well without bothering him again with her flying plans, he

would pay up for a course at a nearby gliding school. This was acceptable to the young woman, who worked hard and kept quiet for six years and then claimed her prize from her father, who was sure she would have forgotten all about it. As in a fairy-tale from the Brothers Grimm, there was yet one more task for her to overcome before Dr Reitsch finally gave in – it was the hardest task of all: a year at college studying to be a German wife and mother. She stuck it out – after that, test-flying a V1 rocket bomb would be child's play – and set off by bike one summer's morning in 1930 for her first gliding lesson.

It's tempting to say that young men are much alike the world over and that the slim, five-foot Hanna would have found any gliding school anywhere in the world an equally tough place to win recognition; tempting but probably, in view of Germany's psychological state in 1930, untrue. Before the First World War Melli Beese had found it almost impossible to maintain her place and right as a pilot; now, dragging the heavy weight of defeat, Communist uprising, inflation and serious concern about their own blond beastliness, German youths were all set to give the newcomer – the only woman at the school – a very hard time indeed. Hanna, knowing she had to prove herself, overcompensated: on her first glide, meant to be no more than the merest hop off a downward slope, she took her craft up too high and landed clumsily, breaking one of the skids. She was given, to the jeers of her classmates, a three-day suspension which, being Hanna, she put to good tactical use, watching the instructors and their pupils until she had their moves down perfectly. When she was allowed back in the cockpit, she performed with such a high degree of natural ability that Wolf Hirth, the chief tutor, took over her training and her fellow pupils accorded her their unstinting respect.

This may not have been unconnected with the political currents swirling around in the atmosphere of the school; the staff and pupils were facing the same choice as everyone else in the country: where did their political faith lie: with the

Communists or the National Socialists? There was no room for the middle way – banners and batons were lifted in the streets, the gutters ran with blood, and the roads ran to the left or to the right; the undecided would be ripped apart. The pupils were all for the rising young politician Adolf Hitler who was offering Germany a new sense of self-belief and certainty in the face of its historical doubt and shame. Hanna had no problems with this at all and, to the boys around her, probably came to exemplify the spirit of young German womanhood braving all odds as she swooped across the hills and valleys of the homeland.

Hirth was beginning to see Hanna as his star pupil and allowed her the privilege of test-flying the new gliders which were coming fast off the production lines of the aircraft manufacturers as they tested aerodynamic possibilities for a future air force. In one of these, a new lightweight training craft, she found herself snatched up by thermals and thrown into the midst of a passing storm, where rain crashed, she later said, like drumbeats against the wings and the controls were utterly useless. Hanna was forced to sit helpless but exhilarated in the cockpit until, in an almost irresistibly metaphorical moment, she found herself bathed in golden light and, looking up, she saw the green earth beneath her. The storm clouds had blown away, the sun was shining and she was serenely flying upside down, achieving a totally new world-view. She took control, flipped the craft and came back down to the airfield to find she had achieved, quite by accident, a gliding height record of almost 10 000 feet. It was a feat that did not go unnoticed by the popular film-makers at Berlin's UFA Studios who were producing a whole series of mountain films with feisty blonde heroines. Hanna was asked to do some stunt flying for a flying film, *Rivals of the Air*, which earned her enough money to take part in a geographical expedition to Argentina, where she put her gliding to use in mapmaking.

Back in Germany, with the Silver Soaring Medal of Argentina proudly pinned to her lapel, Hanna was invited to join an experimental research project, studying the performance of different glider (and later, powered) aircraft designs and materials. She was given a course in powered flight and passed, as usual, with speed and ease; medicine had been long forgotten, she could see a future as a test pilot stretching before her, one which would bring material rewards and the respect of her peers. And it is an interesting point that throughout the early years of aviation, women found it far easier to achieve flying jobs in those states which were not democracies – Russia, Turkey (to an extent) and Germany all provided career opportunities for women. Of course, the penalties for failure were high: German distance flyer Marga von Etzdorf crashed her plane in Aleppo during a Berlin–Australia attempt. She was unhurt and the plane only slightly damaged, but the derision of a French officer so upset her that, rather than bring any more shame on the Reich, she shot herself that night. The new German Chancellor, Adolf Hitler, ordered a state funeral. The message was clear – don't fail, or if you do, burn like Brunnhilde.

Once again, Hanna had no problems with this attitude. Chances were there to take; some of her fellow pilots said she was foolhardy but she didn't agree: she had a brilliant analytical brain and if she looked at the risks and found them acceptable, then she was prepared to go ahead. If they were not acceptable, then, whatever the cost, she would not fly. And she was never lacking in the spirit and courage to tell even the highest authorities that a project was simply not on. In the early years of the Second World War she was asked to test a vast glider, the Gigant, manufactured by the air force for transporting tanks and large numbers of soldiers. The thing needed three aircraft and rocket assists in the wings to get into the air. Hanna realised within moments that it was unstable and refused to test it; the pilot who did, died, as did the pilots of the three towing planes and 120 men aboard the glider.

However, she loved a challenge and when she was asked to fly what was, effectively, the world's first helicopter, the Focke-Sangalis FW61, she happily accepted the job. She wasn't so pleased when she realised she was supposed to fly it indoors, in a vast exhibition hall, after a performance by chorus girls and clowns. The audience weren't that impressed either, as the craft rose up to 60 feet, hovered and flew a slow and precise course. However, as any engineer or pilot could see, it was a stunning example of engineering and flying skill and, where it counted, her expertise was noted.

So was her experimental work on the cutting of barrage balloon wires. With almost appalling simplicity, the boffins had come up with the idea of lining the front edge of an aircraft's wing with a blade sharp enough to cut through steel cable. Hanna was chosen to test the device and did so, flying a twin-engine Dornier. She cut the cable, but one of the ship's engines blew up. She managed to nurse her plane to a nearby airfield, not knowing that the whole test had been observed by the air force's Chief Technical Officer, First World War flying ace Ernst Udet, who was on his way to see Hitler. He was an old friend, and possibly one-time lover, of Hanna and told the Führer all about the fearless aviatrix. Unfortunately for Udet, the Führer's court was a nest of jealousies and he got caught up in Air Supremo Herman Goering's embarrassing retreat after his planes' failure to clear the skies over England in the summer of 1940; Goering was trying to shuffle the blame on to his old flying buddy and Udet shot himself, depriving Hanna of a friend at court.

So, in typical fashion, she went to Hitler himself and charmed him with her vivacity, her stories and her devotion to the cause. She wanted, she told him, to test the toughest planes the air force could come up with; he told her to go and see the men at Project X; she did and they came up with all she had wanted, and more!

The DFS rocket plane was a killer – the last in a line of killers – experimental craft that didn't make it; but nothing ventured and, realising the huge advantage that rocket speed would give to fighters in aerial combat, the researchers at Project X kept on building new models, each one better than the last, until with DFS 194 they had a plane that could fly and was reasonably manoeuvrable. The 194 became the Me163 which was tested during 1941 at speeds approaching the sound barrier. Here, all sorts of problems began to make themselves apparent and, when the chief test pilot was injured during a flight, Hanna was next in line. Many of the male pilots at Project X thought she was using her influence to get a job her experience didn't qualify her to do: they called her a star who had to be number one. Maybe she was – after all, the Me163 had smashed every air-speed record in the world (not that the world was particularly interested in speed records in 1941) and the chance to be part of history, after the Reich's inevitable victory, was not to be overlooked. Nor was the risk: that was very real too.

On Hanna's first test flight, after the rocket plane was towed into the air, the detachable undercarriage would not detach; she knew she couldn't land with full fuel tanks so had to cut her tow and fly in circles while the plane bucked and shuddered, seriously unbalanced by the weight of the undercarriage, until she'd run her tanks low; then she landed, missing the airfield, hitting a ploughed field and smashing the bones in the front of her face on the instrument panel. She took the time to write up an accident report before attending to her own bloody wounds. She knew, as did the pilots who watched the flight, that she could have bailed out at the first sign of trouble, leaving the rocket plane to crash; it's what any ordinary pilot would have done, but not what was expected of a test pilot of the Reich. Meeting Goering after her recovery, Hanna offended the air chief by telling him bluntly that the Me163 would never go into mass production. No friend of Ernst Udet, Goering was now no friend of Hanna Reitsch.

In February 1944 Hanna was personally awarded the Iron Cross for her heroic work as a test pilot by Adolf Hitler at his Berchestgarden retreat. Hanna had more on her mind than the honour she was receiving; she was thinking of suicide. The V1 pilotless bomb had been tested and was in production and would soon be launched at London. Hanna's original idea had been to convert the flying bomb into a piloted craft with minimal armaments but the same explosive load so it could be guided to selected targets rather than fall haphazardly to earth within a given area. The pilot would be provided with a parachute and could jump from the plane before impact – although, as anyone who looked at the design of the plane would see at once, the large rear-mounted jet motor would hit the cockpit cover and/or the pilot as soon as he or she climbed out, thus the craft was, effectively, a suicide plane. Hanna didn't seem to have any problems with the idea of dying for the Reich, and was sure she could recruit a squadron of like-minded heroes; so certain was she that she started putting out feelers and receiving positive answers. Hitler was not keen – he still had hopes that Germany could pull a victory out of the hat somehow; he was also concerned about the effect on public morale once people heard that the country's best pilots were actually in the process of committing suicide, no matter how the whole business was being spun by the propaganda ministry. He rejected the idea, telling Hanna that the air war would be won when the new jet fighter from Messerschmidt came on line. Hanna disagreed, there and then – in itself an almost suicidal move at this period in Hitler's decline into mania. She disagreed and told him he was talking nonsense; as a test pilot, she knew the jet was too late. As someone with common sense, she should have known that the suicide plane was also pointless, but it was her idea and she had never lacked confidence in herself. The Führer kept his temper and, like Dr Willi Reitsch, told Hanna she could follow up the idea as long as she didn't bother him with it.

Goering thought the idea a non-starter, as did his top commanders; as a group, German air-force pilots weren't any more fanatical about Nazism than they were about suicide. However, Hanna found a supporter in SS Colonel Otto Skorzeny, noted for his commando-style rescue of the discredited Mussolini; together, with Hitler's tacit approval, the pair went to the V1 development site at Peenamuenda where engineers set about turning a number of flying bombs into planes. The craft were to be launched in the air by being dropped from a bomber and although the experimental models achieved both separation and level flight of a sort (the ride was unimaginably bumpy) attempts at landing (without an undercarriage) were inevitably fatal to the test pilots. Hanna and Skorzeny were not put off; the planes could be patched up and they were not likely to run out of pilots. Eventually, however, the authorities ran out of patience and called a halt to the testing. Hanna wasn't impressed by Skorzeny's obedience to his orders. She told him that she had always thought he was the kind of hero who wasn't afraid to disobey a cowardly command. She demanded, like Brunnhilde, the right to do what the Leader really wanted; she was going to test the plane herself.

Skorzeny gave in and she went up, secured in the tiny and highly unstable craft beneath the belly of a bomber. It was a ridiculous situation. It has been said often enough that by late 1944 the German High Command was living in a dream world that had lost touch with any kind of reality, and it is equally clear that Hanna and Skorzeny and their team were full-time subscribers to the Nazi Club Mad; and Hanna, at least, was fortunate to escape with her life. After a first successful test of the Riechenberg R11, as the piloted bomb was known, she suffered two crashes, sustaining injuries each time, but never quite enough to put her out of the game. She would not give up her dream; modifications were introduced, more flights were made and slowly the piloted bomb began to look as if it might work. A cadre of pilots was gathered, an operational programme set out and then the Allies invaded Europe, advancing at such a speed that the proposed airfields – within the crafts' fuel range of London – were overrun. The project was closed down.

Soon, even true believers like Hanna had to admit that the war was going badly for Germany. Driven back on two fronts, Hitler was battened down in his Berlin bunker as the city was smashed to pieces all around him by the artillery and bombs of the Russian army. As George Steiner once said, this was the last act of Götterdämmerung but for real and with no Rhine to overflow and put out the fires which raged non-stop; the Third Reich was going down in a blaze of destruction and Hitler decided he wanted to see Air Force

General Roitter von Greim at his headquarters. It was, of course, impossible to get into or out of Berlin, a city under siege, but the impossible had never bothered the Führer before and it didn't now. Von Greim realised that if he was going to answer the call – and a number of his fellow generals had already baled out and refused to go – he would have to fly in; he was an accomplished pilot himself but decided that this was a mission which required back-up and he called on his old friend and sometime lover, Hanna Reitsch to be his co-pilot.

The two were flown by an air force pilot to an airport on the outskirts of Berlin; they were lucky to get that far. Russian fighters knocked out half their escort and peppered their plane with bullet holes. Here they transferred to a light Fieseler Storch two-seater and, with Rheim piloting and Hanna handling the navigation – since the smashed city streets below them were totally unrecognisable – they took off and headed for the centre of Berlin. They flew low, hoping to escape the constant enemy fighter patrols; this, of course, meant they were vulnerable to ground fire and the Russians let loose with everything they had at the small plane. Greim was hit by exploding shrapnel and as the Fieseler spun down towards the rubble, Hanna threw herself across his body and grabbed the controls. She was unable to use the rudders but, on stick alone, she pulled the craft out of its dive and, flying through a storm of steel far thicker and more violent than the thunder which had snatched her glider all those years before, she made for the landmark of the Brandenburg Gate, where she managed to land without hitting either shell holes or fallen masonry. A passing truck was commandeered and von Greim driven to the bunker where the occupants were preparing to go down with the Reich. Whatever Hitler had wanted to see Greim about was no longer relevant and, always a man for the personal touch, the Führer presented Hanna and von Greim with their very own suicide pills. Done out of her chance to commit suicide for the Reich by airborne bomb, Hanna was proud, at last, to be able to perish thus at the side of Germany's greatest leader.

It was not to be; hearing of moves by SS Chief Himmler to negotiate a peace with the Allies (they were all, indeed, living in a mad world) Hitler decided to send von Greim and Hanna out of Berlin with orders to stop any talk of surrender. The German people had proved unworthy of his great vision, so now they must pay the price and perish! Hanna protested – she wanted to stay and die – but orders were orders, and she and von Griem made their way back to the Brandenburg Gate where an air force pilot picked them up and flew them out of the city.

The end was inevitable, but took its time in arriving – there was room for more tragedy to come. As the nation crumbled, Hanna's father, Dr Willi Reitsch, emulated his beloved Führer and killed himself, his wife, his younger daughter and her two children, and the family maid; Hanna did not, in her autobiography, appear to feel this was in any way reprehensible; rather, it was a shouldering of the ultimate responsibility by the head of the household. Not surprisingly, it took a year and a half before the Allies considered Hanna sufficiently de-Nazified to be let loose in the post-war world.

She went back to her first love (gliding rather than God, the missionary idea seems to have got lost somewhere along the way) and with her undoubted skills and experience began to earn a good living testing new craft. However, she was missing something of the old magic and, after a disappointment when she was refused a visa to travel to Poland (no great love of ex-Nazi test pilots there) she moved her operations to Ghana. Here she found a new leader in the shape of a charismatic African dictator and, as she titled her last book, *Flew for Kwame Nkrumah*. She died, not by suicide but as the result of a heart attack, in 1979.

8
Always stand on the right

*T*he stagecoach had left; there was no way out of town. The snow-topped mountains brooded against the darkening blue sky of evening. On Main Street a few kids were playing but nervous parents soon called them in. A lot of the shops were shutting up early today, but no one was going home. They were waiting for the showdown, every eye peering at the closed door of the town newspaper, *The Bulletin*, every mind wondering what the crusading young editor – at 24 he was hardly more than a kid himself – was thinking now about his editorials exposing local corruption. Maybe he wasn't thinking about them at all, maybe he was composing his own obituary; he had every reason to: Jack Culpepper down at the saloon had promised to shoot the dirty sidewinder full of holes if he dared show his lying face on Main Street.

It was so quiet you could hear the handle on the newspaper office door turn, the hinges squeak as it opened, the sound of the editor's boot heels on the wooden sidewalk as he stepped out into the chill evening air. Like everybody else in town, he knew it was a case of showing up or of being shown up as a man who lacked the courage to stand behind his words. Slowly, keeping his hands well clear of his body, his jacket open to show that he carried no weapons, the young editor began to walk along the creaking wooden boards towards the saloon. Behind him, the sheriff walked casually, his hands in his pockets but his reputation as a quick-draw sharpshooter as loud as a shot in the silence. The two men, the younger well in advance, walked into the light of the evening sun, their shadows lengthening behind them. And the silence went on – no sudden crack of gunfire echoed around the mountains; the editor kept walking, getting closer and closer to the saloon with every step, until he stood under its painted sign and, after pausing for a moment, walked on through the swing doors and vanished inside.

Jack Culpepper watched him as he crossed the wooden floor to the bar and placed a newspaper flat on the polished mahogany surface, so Jack couldn't avoid seeing the front page. It was a story about a murder, a slight case of homicide, as the headline blared, and it told how a desperado named Jack Culpepper had killed a man in Montana and spent time in the state penitentiary for doing so. It was a story Jack thought had been forgotten; a conviction that could get in the way of his ambitions as a saloon owner. It was a story, the young Editor told him, that hadn't been printed yet; this was the only copy, but it could be set on the new press any time and within hours thousands of copies would be run off and distributed as far the state capital. It was up to Jack.

He was a man of sense. And became, that same night, a stout supporter of the young fellow at *The Bulletin*. The story never appeared, and after a year or so the editor moved on to other things and other places but one thing he never forgot was the power of publicity.

At about this time, 1912, in the fishing port of Hull in the north of England, a nine-year-old girl called Amy was shepherding her younger sister to the school they both attended. It wasn't a very good school; the curriculum was a mixture of all sorts of subjects, none of them taught really well and Amy was bored for a lot of the time. She preferred being at home in the evenings, where her mother would play the piano and the girls would sing along with comic songs. Later, their father would come back from the family fish business and, as often as not, play trains with Amy.

When she was 12, after staying an extra year in junior school to keep her sister company, she moved on to the local secondary school where she was placed in a class a year below her natural age. She had no problems with the work and because she could finish it faster than most of her classmates, had more time for mischief. However, she never thought of it as such; she had, even at an early age, a well developed sense of fairness and there were a lot of things she saw around her that didn't measure up to her criteria. Like swimming lessons, for a start. There weren't any because the headmaster did not feel that this was a proper sport for young women.

Amy set to work undermining the institution: she persuaded her father to hire the local swimming baths, she managed to inveigle some of her teachers into escorting girls for swimming lessons and, somehow, contrived to keep the whole exercise quiet enough to avoid the head's notice. She was also instrumental in organising a school-wide protest over the uniform hats the girls had to wear whenever they left school grounds. This time, she learned another valuable lesson: when she arrived on the morning of the protest wearing alternative headgear she found that, contrary to the agreement worked out among the girls, no one else had swapped their hats. She was hauled up before the head, given a stern talking to and suffered the sort of ostracism that often descends upon those who make the rest of us feel somehow inadequate or cowardly.

She became less outgoing, not only because of the hat business but due to an accident on the sports field, when she was hit in the mouth by a cricket ball, a viciously hard missile when travelling at anything more than a snail's pace. Her front teeth were damaged and the local dentist was unable to rectify the problem; the school authorities might well have pointed out that this is what happened to girls who tried to play boys' games. For the next few years Amy kept her mouth shut as much as possible and cultivated a closed-mouth smile, which gave her a mysterious and knowing air.

And as she got older, 'knowing' is what she wanted to be. She was never happy being a child – though not unhappy *as* a child – she wanted to grow up and be sophisticated and, as anyone who went to the pictures knew, to do that you needed a man. She found him when she was 18; his name was Hans Aregga, he was a Swiss businessman and he was 27 years old. In his official biography, *Amy Johnson, Enigma in the Sky*, David Luff, the first writer to chronicle the affair in detail, follows the ups and downs of the two as they circled each other (Hans concerned about Amy's youth at first) then became lovers when she went to Sheffield University; became unofficially and then officially engaged after she left – there were visits to his family and hers – and finally, as Amy began to realise that he had never intended to marry her, drifted apart with the inevitable pain. Luff makes it plain that Amy was the innocent party in the relationship, her affections and naivety used and abused by Aregga who didn't really want to commit to this young woman but couldn't quite bring himself to let such a devoted and sensual treat go. Her letters are passionate, his are Swiss; she talks about the future and the stars, he talks about business and his family; she climbs mountains and throws herself off the peaks; he turns the Alps into Holland. In many ways, he was exactly what she was looking for: a real affair with passion and heartbreak on her part and a supremely boring temperament on his. Years later, when she was on top of the world, she would take a chance with the real thing and find the way out of that adventure was by no means as easy or as painless.

A young woman with a degree, even in economics, was unusual but by no means unknown in the early 1920s; the old world was changing fast under the influence of the new media; records were comparatively cheap, so was the cinema; weekly and monthly magazines were growing in number and though the newspaper business was still conservative, the more popular section of it was learning fast how to follow the story. And Amy was reading stories and seeing films and finding the world they depicted a good deal more interesting than the prospect of settling down to find a job. She moved around the country trying various things: she worked as a trainee for the John Lewis department store chain, she learnt shorthand and typing and tried secretarial work in a lawyer's office; it was all boring, boring, boring. Once she'd got Hans out of her system, the prospect of settling down with someone else seemed no more attractive than slaving away in an office or on a shop floor. She was looking for an adventure and, all the time, it was flying right over her head.

She'd moved to London and was sharing a flat near Edgware. During the summer weekends, when she was out in the garden or playing tennis at local courts, planes from the nearby Stag Lane airfield, where the London Aeroplane Club had its hangars, passed over the houses in what to most residents seemed like an endless procession of noise. The dreaming suburbs didn't like to be disturbed on their days off; Amy, on the other hand, hating her weekday work, found the spectacle more interesting. She had flown once before, a five-shilling trip round the airfield at a Hull show, but had not been that impressed. Now she began to think again: she was in her twenties, she was stuck in a boring job with no prospects, there was nothing she wanted to do with her degree; her family was still in the north so she had no personal responsibilities: why not fly? There was no great revelation, no leap to flight; there was, however, a feeling that this would be something different that would demand her total commitment.

A few weeks later, in August 1928, she caught a bus to the Stag Lane airfield, where she found that flying lessons cost £2 per hour and the club tutors reckoned most people would need from eight to twelve sessions. She wanted to sign up there and then, but had to wait a few weeks (how very English) to be elected to club membership. Once it came through, she began to learn to fly.

Meanwhile, in New York, George Putnam, ex-*Bulletin* editor, now publisher of true-life adventures like Charles Lindbergh's *We* and the story of Commander Richard E. Byrd's flight over the North Pole, was sitting down in the office of attorney David Layman. What Putnam wanted was to worm a secret out of the lawyer. What the lawyer wanted was a young, attractive, respectable American girl.

A few months before, rumours had reached the publisher of a new attempt to fly the Atlantic from west to east; as far as he could ascertain, a rich American woman had bought a plane, hired pilot and navigator, and was all set to be the first woman passenger to cross the ocean. There was, surely, a book in it and he was certainly the man to publish it, if only he could find out more. He set a friend, reporter Hilton Railey, to track down the facts; it didn't take long. Mrs Amy Guest, a rich socialite who had always had a hankering to be a woman of action, had bought the plane from Commander Richard Byrd and named it *Friendship*. However, once her family found out about her harebrained scheme, they said no. Denied her chance, Mrs Guest still wanted to ensure that the first woman to cross the Atlantic would be an American. She was willing to underwrite the attempt so long as the right kind of girl could be found.

The publisher wanted to know what that meant – he also wanted to know if Putnam's could handle the book of the flight. Layman told him there was no problem with the book, just as long as the right passenger could be found. There had been a possible candidate: Mabel 'Mibbs' Boll, known as the Queen of Diamonds, on account of the vast amount of jewellery she always wore; perhaps Mrs Guest was worried that all that glitter would weigh down the plane or it could be that she just didn't want to be associated with anyone that vulgar – she had, after all, married a duke. Time was also a factor. Just six months earlier, aviatrix Ruth Elder had attempted to fly her own three-motor Stinson across the ocean; she had come down near the Azores and was rescued by the Navy; she would surely be trying again soon.

Putnam agreed; his instinct told him that an Atlantic crossing by a woman as pilot or passenger was going to happen – it was just a question of who was there to garner the benefit of the publicity. After more negotiations, Layman agreed in his turn that the publisher should try to find a candidate. Putnam discussed the problem with Railey, who had recently spoken to an old naval friend who had mentioned a young social worker of his acquaintance, working at a reception centre for immigrants in Boston. She was everything Mrs Guest could want and, what's more, she had a pilot's licence as well; her name was Amelia Earhart.

Amelia was born in 1897 and spent her first 11 years in Kansas City. Her father, Edwin, was a claims assessor for the Rock Island Line Railway; he was something of a visionary as well, and liked to tinker with inventions. He also liked a drink or two but, at least, during Amelia's childhood, it was a strictly social practice and no bar on his steady climb through the ranks of the railroad until he was sufficiently senior to have a railway carriage for his own personal use. Amelia and her younger sister Muriel attended local private schools and, helped by the family position and money, Amelia had few problems following her natural bent towards woodwork and mechanics rather than housework and dolls. Her mother, Amy, had a questing intelligence and when she prepared chickens or ducks for table, would virtually dissect them, taking her fascinated elder daughter through the inner organs of whatever animal they were having for dinner that night. She also made special playsuits for her children which freed them from the cumbersome and restricting dresses and petticoats of the time and allowed them to run and climb trees. Amelia's father also encouraged her tomboy nature (a Christmas letter to him asked for footballs, since the sisters had enough baseballs, bats etc.) and, in what seems almost a ritual for American fathers of independent daughters, gave her a rifle so she could go out and kill small mammals; in her case, rats which were infesting a family barn.

What she couldn't do anything about was her father's drinking, which people began to notice around about 1911. He began to make mistakes at the office, started arriving home late and his usually calm temper got worse whenever he'd had a few. The company supported him; he was a good lawyer, but in a spirit of what might be taken as self-destruction, he rejected the help, the drinking got worse and he lost his job. The family moved to Des Moines, Iowa, as he searched for increasingly less responsible and less well paid work. Amelia attempted to remonstrate with her father on a number of occasions and once, finding a bottle of his whiskey, tipped it down the sink. He was so furious he would have hit her, had not his wife stepped between them. It was a deciding moment. Edwin and Amy would separate, not immediately, but the bonds that had existed between them were broken and the two would drift apart.

For Amelia and Muriel, the loss of earnings meant no more private schools, but so dissatisfied with public education was Amelia (and she made no secret of her feelings, informing the school authorities whenever she felt a teacher was inadequate) that Amy cashed in a family trust fund and sent her elder daughter back into the private sector.

After graduation, Amelia decided to study medicine but her plans, like so many, were changed by America's entry into the First World War in Europe. She was visiting her sister in Canada when she couldn't help noticing the walking wounded all around her; men who had returned from the carnage in France minus one or many limbs, often with other disfiguring wounds. One day she saw four one-legged ex-soldiers walking as well as they could down a road and her feeling crystallised into a desperate need to 'do something'. Thinking about what she *could* do, she settled on nursing and had no hesitation in setting aside her university career to train and go to France.

Once overseas, Amelia showed a self-destructive need to work harder, achieve more and rest less than was strictly sensible. During the flu epidemic that raged around the world during 1918 she was more or less in sole charge of her wards and, not surprisingly in her exhausted condition, when she contracted the virus herself, she was in no condition to fight it. She was returned to America where she spent some months recuperating. The general joy at the news of the armistice did not communicate itself to the young woman: she felt there was a total lack of thanksgiving and an equal lack of the understanding that Germany had learnt nothing by this defeat. She was, undoubtedly, perceptive, compassionate, serious, forward-looking and one can quite see why her friends must sometimes have wanted to shout at her: relax, have some fun. But she was not a fun person; she enjoyed herself and could be good company but there was always something calculating about her attitude, rigid about her thinking, as if everything should, at some level or other, be explainable and therefore, improvable.

Fully recovered and cheered by the news that her father had not only given up drinking but had opened a law office in California, Amelia and her sister decided to join him in the sun – they also had hopes that their parents might get back together again and Amelia had no qualms about manipulating the situation to what she could quite clearly see would be everybody's advantage.

Once settled with Edwin, the girls started looking around with a view to finishing their interrupted education. However, Amelia's eye fell first on a local airstrip, Rogers Field, where flying lessons were available, a course costing $1000. She'd seen and become interested in army planes in Canada and France and now decided to try the experience for herself. Her first short flight left her convinced. She ought to fly – she would fly. This was what she wanted. She hurried home, announced her wishes, asked her father to stump up the $1000 and was amazed when he wouldn't, because he couldn't, even in the face of his daughter's utter certainty. Unabashed, she got a job with the telephone company – one somehow can't imagine the personnel officer turning her down – and earned enough to pay for her own lessons with Anita Snook, a local teacher. She also completed the cutting process that had been shortening her waist-length hair inch by inch over the past year or so, and emerged from the local barbers with a bob; so much more convenient under a helmet. She would cut it back further in years to come, but for now, the bob was quite radical enough for her family to cope with.

She was not and would never be a natural pilot. In the air, as never on the ground, she was careless, as if flying freed her from the often obsessive concerns she showed in her everyday life. Not stupidly careless, or dangerously careless, but no pilot ever is until the stupid or dangerous moment. Her instructor told her she should learn to fly by feeling but since she didn't live by feeling, why would she suddenly discover the ability to do it once she left the ground? But good or mediocre, she flew and did her solo and got her licence. To celebrate, she bought herself a full-length leather flying coat and carefully distressed it with oil and boot heels so it didn't look too new. She also persuaded her mother to help her buy her own plane, a Kinner biplane, which she took up every chance she got, and took up high enough on one occasion to gain an altitude record of 14 000 feet.

In the meantime, there was the material question; she didn't consider earning a living through flying. She wasn't a barnstormer, she didn't have her own money. She flew, as she said, for the fun of it, although one wonders if the statement – the title of her second book – is a little disingenuous. She enjoyed it as much and probably more than she enjoyed anything in her life, but there was nothing in her of the sudden storms of enthusiasm or despair that battered Amy Johnson's inner cloudscape.

Denison House was a receiving and education centre for poor immigrant families and Amelia applied there for a position as novice social worker. She got the job and was superintending language classes when she got a call from a reporter named Hilton Railey who wondered if they could meet. She agreed, suspiciously, but learned little beyond the fact there was a project that involved flying and a further interview in New York with the publisher G. P. Putnam. Railey, with the sure eye of a newspaperman, immediately spotted Amelia's resemblance to Atlantic flyer and *bona fide* world legend, Charles Lindbergh. She could have been his sister. He immediately coined the nickname Lady Lindy; or so he said. George Putnam also claimed to have invented the name. Either way, one of them did and kept it in reserve until the moment it would be most effective.

Amelia went to meet Putnam in Manhattan and was less than impressed by his casual attitude. He was most impressed by her, both as a person and potential publicity, and by the end of the meeting decided she would be the one. A couple of days later he made an offer and she accepted. She wasn't that keen on the idea of being a passenger, although she accepted the position since she knew her experience didn't include three motor planes or covering long distances by instrument flying. On the other hand, given Mrs Guest's intentions, it was vital that the American girl should be seen as an important part of the whole mission. A contract was drawn up between the pilots, Wilmer Stulz and Slim Gordon, both highly experienced flyers, stating that Miss Earhart was the captain of the craft, representing the owners and that no decisions should be taken without her agreement. In turn, she was enjoined to keep the trip secret until the day of take-off, to avoid any last-minute competitor beating Team Guest to the prize. Amelia went back to working at Denison House while Stulz and Gordon worked on the float plane moored in Boston harbour. The main thing was to keep Amelia away from the scene, so she only got one glimpse of the golden wings and red-gold body of the *Friendship* before she climbed aboard for the flight. She was as aware as anyone that the colour was not for looks alone – it was adopted because in the case of a sea landing the colour would be more likely to show up; not that likely in the middle of the Atlantic, but there would be at least some chance of rescue.

The *Friendship* took off from Boston harbour on 3 June 1928. The first leg of the flight was to Trepassey, Newfoundland, which would provide the closest jumping-off point for a landing on the west coast of Ireland. Unfortunately, once they touched down, the weather closed in and they were unable to start for two weeks. At one point, George Putnam cabled Amelia: 'Retire with nuns and have your laundering done.' She cabled back, 'Socks underwear worn out. Shirt lost to Slim at rummy.' She signed the message AE, a form she was to use exclusively from then on.

The *Friendship* took off at last on 17 June and headed out over the Atlantic. The three 250 hp Wright Whirlwind engines functioned perfectly and so they should; Slim Gordon was one of the most experienced mechanics flying. There was no problem with the piloting of the ship either, since Stulz was as brilliant a pilot as Gordon was engineer; the problem lay in the man and his drinking habits. This was history repeating itself, as AE crouched behind the pilots in her tiny passenger space and saw the whisky sloshing around in Stolz's standby bottle; she only hoped he didn't start to need it. As it was, he took off with a pounding hangover from a session the night before. But as the flight proceeded, so his headache lifted and, concentrating on his instruments, he didn't need the stimulation of alcohol; even so, AE seriously considered emptying it overboard.

The flight took just over 20 hours and proceeded with little difficulty. The weather stayed fine, though not clear, they were flying through the overcast for most of the trip; Stulz stayed sober; Ireland, however, had not stayed where it should be and when they came in to land they found themselves bobbing in the harbour of the Welsh town of Burry Port. It was an indication, if any were needed, of how even the most experienced pilots could misjudge distance and heading when flying by instruments in overcast conditions.

Wires were sent to Putnam in New York, to Mrs Guest in London and Hilton Railey in Southampton. He flew down immediately after cautioning Lady Lindy not to say a word to the press; her story was already bought, if not yet paid for. It wasn't easy; after an initial bemused lull, the Welsh welcomed their visitors with gusto and representatives of every newspaper with an office in London descended on the little seaside town. The effect was everything George Putnam could have wanted; the press loved Lady Lindy almost as much as the camera did. With her tall, slim figure, her open, pretty but not challengingly beautiful, face, and her modest manner she was made for the modern media and they were determined to make her a star. And though she wasn't expecting any of it, and insisted time and time again that she was only the passenger, that Stulz and Gordon had done the flying, it didn't make any difference because George Putnam knew exactly what was going to happen and how he intended to capitalise on it.

The flyers transferred by air to Southampton (leaving Burry Port just as Sir Arthur Brown of Alcock and Brown, the first flyers to conquer the Atlantic, arrived to congratulate her) where Mrs Guest welcomed her American girl at the quayside. The party then travelled up to London where thousands of spectators struggled to get a glimpse of Lady Lindy and no one at all took any notice of Stulz and Gordon. Not that they minded – they did the sights and visited a number of airfields, where they got to chat with other pilots who understood exactly what they'd done. Wilmer Stulz was also able to drink his way across the country without causing himself or Mrs Guest any embarrassment.

The spotlight was firmly turned upon AE who had to endure endless receptions, crowds, banquets, nightclubs and, most annoyingly, critical comments from the classier newspapers, who took themselves too seriously to give in to the general delirium without making the object of their obsession pay for her popularity. The *Evening Standard* said that Miss Earhart's presence contributed no more to the flight than the presence of a sheep. C. G. Grey, in *The Aeroplane*, compared AE's crossing unfavourably with recent fights by Ladies Heath and Bailey. And AE agreed with every word of it; she knew she hadn't flown the distance, couldn't fly it yet but she also knew that she would make sure in the months and years ahead that she would gain the experience to make her own record attempts. And perhaps she was aware that the man who would help make her dreams possible was heading towards her, sailing across the Atlantic (Putnam hated flying) as fast as his luxury liner could carry him.

While all this was going on, Amy Johnson was getting her hands dirty at Stag Lane. The flying lessons hadn't got off to a good start: her first instructor told her she'd never learn anything if she didn't listen. This was due to the fact that she couldn't hear – she'd been given a flying helmet several sizes too large, so the earphones were in the wrong position. It probably also had something to do with the instructor's attitude. The club did have a few women members but none of them with a northern accent and so decidedly lower-middle class. However, once she moved on to another instructor, things improved and she began to get some miles into her logbook. She also started arriving early at the Aero Club so she could spend time in the workshops observing the mechanics as they stripped, rebuilt and tuned engines for the club members.

Jack Humphreys, the chief mechanic, struck by Amy's interest and hard work, was happy to let her become a sort of unofficial apprentice; the other mechanics in the shop started calling her Johnnie, which took care of the unfortunate fact that she was a girl learning about engines, and within a few weeks she became part of the team, learning invaluable lessons about maintenance and a few hints about the class struggle in north London. The club President called for Miss Johnson to be expelled from the workshops, the airfield, the club itself, for fraternising with the other ranks; Jack Humphreys promptly informed him that if Miss Johnson walked, the workshop would walk with her: the protest was quietly dropped and Miss Johnson went on acting like a lady shouldn't.

She soloed in June 1929, gaining her A licence, and in November got her engineering C licence, the first woman to do so since Lady Mary Heath, who had also learned to fly at Stag Lane. She was, by all reports, a competent but not a brilliant flyer. She worked hard to gain her skills, harder still to hone them, but it was never a question of flying by feel or instinct and, throughout her career, her landings and take-offs were a matter of concern. She mastered the technique through practice and effort, because she wanted to fly; she wanted to fly because she saw it as a way of making her mark on the shifting world all around her – if the mark was clear enough, real enough, then she too would be real; unlike her younger sister Irene who committed suicide in 1929, shortly after getting married. The marriage, as far as anyone knew, was happy; there were no emotional or financial problems: everything seemed set for a happy, untroubled married life. Did the sisters share a temperament? Did Irene experience the same highs and lows that switchbacked Amy's emotional life and always threatened to take her either too high or too low? And to what extent did Amy's decision, taken soon after the suicide, to fly solo to Australia, when by any standards she was totally unprepared, have to do with her own understanding of what sitting quietly in an English garden might do to her? She was not by nature given to content; she lived, one suspects, with that old trouper of the anxious mind, Nameless Dread, and had learned, as Irene may not have done, that keeping busy is the best medicine and that flying to Australia was about as busy as a body could be.

One side-effect of Irene's death was her parents' added support and concern for their elder daughter. The family had always been close and now, when she began to broach her plans, they overcame their natural concern and supported her all the way, going so far as to put up half the price of a plane. The other half came from Lord Wakefield, a philanthropist, who fell under the spell of Amy's enthusiasm. This was just as well; the press weren't interested in an unknown young woman (she obviously lacked Chubbie Miller's persuasive talents) and she was advised by an Australian High Commission official to go by steerage rather than by air. However, Wakefield's money and his logistic support (his company sponsored the oil and petrol) made it possible for her to buy a second-hand Gypsy Moth with a 100 hp Gypsy engine. Its top speed was 98 mph. Jack Humphreys immediately got to work reconditioning the engine while the plane was repainted and named *Jason*, after the Johnson family fish business (which, to be fair, *was* named after the hero). Amy took a crash course in navigation and got as much advice as she could about flying in the many different weather conditions she would be facing during the journey. Her aim was to cut the record time set for the journey in 1928 by Bert Hinkler and, by careful planning of the route, she reckoned she could do it. Just about everybody else, apart from Jack Humphreys, reckoned she couldn't and shouldn't: it was the kind of thing chaps did, or Americans. If Englishwomen had to try, then let them be ladies rather than a fishmonger's daughter from Hull.

She set off from Croydon on 5 May 1930. Loaded with extra fuel tanks, the Gypsy Moth flew heavy and she had to make a second run before getting into the air. There was only a small crowd to see her off; the flight simply wasn't newsworthy. She headed out towards the Continent, landing first in Vienna, where the ground crew wouldn't let her touch her own engine, and then in Istanbul. She was keeping to time, taking 48 hours so far, and the only serious wear and tear was on her right arm, from the endless pumping needed to transfer fuel from the storage to the engine tanks. The third day took her over the Atlas Mountains and into thick cloud. She went down, trying to get under it, and found herself flying into a tremendous wall of rock. She banked as hard as she could and skimmed the mountain wall by inches. After that, she found a nice safe railway line to follow into Aleppo. Her first job on landing was always to see to her engine, checking everything each night, a task that took up to three hours and left little time for sleeping, since she always started as early as possible.

The next leg was a 500-mile desert hop to Baghdad. Half-way there she ran into a sandstorm. Her goggles became clogged, and the engine began to stutter, so she took the plane down, landing virtually blind, clambered out and anchored the undercarriage as best she could, covered the engine cowlings, then sat on the tail, to stop the craft from being blown over. All around her, through the swirling sand, she could hear the howling of the wild dogs which, she'd been told, would attack and tear travellers to pieces in minutes; she pulled out her pistol and waited for them to attack. It was a long way from England.

Eventually the storm abated, the dogs slunk away and the plane, miraculously, took off. On landing at Baghdad, one of the wheel struts gave way but fortunately there was an Imperial Airlines station handy and their technicians were able to supply spare parts and recondition her engine, allowing Amy an early night for once.

Next it was Baghdad to Bandar Abbas, then on to Karachi, cutting two days off Bert Hinkler's journey time and establishing a new Britain–India record. Now the newspapers were beginning to take notice – just when her luck ran out. On the leg to Karachi, she was held up by a stiff head wind and, running out of fuel, put down near a military post where she ran into an obstruction and damaged a wing. Local tradesmen got to work and soon had the wing repaired but she was losing time and was to lose even more on the next leg, when she landed in a field outside Rangoon. She'd been unable to find the racecourse which served the city as an airfield, and upended the Moth into a ditch. This time the students of a local engineering school turned out to help and repairs were effected – though she was now well behind Hinkler.

By the time she reached Singapore, she was even further adrift, but that didn't seem to matter to the crowds who waited to see her land. It was slowly becoming apparent that the indifference that had greeted her departure was changing into something else. Her father wired her from England, sending love and congratulations but also laying out some of the many business offers that were flooding in and that he was now handling on Amy's part. It began to look like every mile she covered increased her fame by a margin of ten.

Flying down through the Dutch East Indies was made even trickier by the lack of proper maps, each of which seemed to be slightly different; however, the biggest setback came when she was forced down by a violent rainstorm and landed in a field where she ran over a series of building poles, ripping the fabric of her lower wings. The district was scoured for tape and plasters and enough were found to make good the damage. More bad weather added extra hours, but there was no chance of beating Hinkler by now and Amy had settled on just reaching Australia as her goal. By 23 May she was at Attemboa, the jump-off point for the 500-mile flight across the Timor Sea. Whether it was engine maintenance or consideration that caused her to wait until 24 May, Empire Day, before setting off on the last leg is a moot point but it was a brilliant decision, underlining her arrival in Darwin with all the majesty of the British Empire celebrating its own great traditions.

She was instantly famous. The typist who flew to Australia; the office girl who quit her dull job to find adventure; Amy, Wonderful Amy; the Empire's great little woman. They loved her. In Australia and back in Britain, they loved her because she was like them, because she was ordinary; no Lady This or the Duchess of That or Mrs the Hon. Double-Barrelled-Cut-Glass-Accent; she was just plain Amy, truly a girl of the people, of the mechanics back at Stag Lane who called her Johnnie.

Charles and Ann Morrow Lindbergh sent a telegram of congratulations; so did the King and Queen, and they threw in an OBE to boot. The *Daily Mail* offered £10 000 for her story and her plane. She'd done what she wanted to do, and it was exhilarating. She realised that, more than anything else, she wanted to stay there, to do more, to fly out of the overcast, above the clouds into the sunlight.

AE and George Putnam got married on 7 February 1931. The night before the wedding, AE had written a letter to George in which she stated her own reluctance to marry and said she would not expect him to be faithful any more than she expected to be so herself. She asked that neither should interfere with the other's work or play, and said that if, a year after the date of the wedding they had not found happiness, they would let each other go.

After the Atlantic crossing and the hullabaloo in London, AE was glad of the chance to find a little peace and quiet travelling back to America with George; once there, Putnam set about creating the Amelia Industry he'd always had in mind. He was far from a fool and realised from the start that there was only so much money to go around the small number of women flyers in the country and that if AE wanted to stay in the business, they needed to corner the market. First there was the book, *20 Hrs, 40 Min: Our Flight on the* Friendship, which had to be in the bookshops straight away to capitalise on the lecture tour. He gave AE advice on delivery, content, style, suspense, arrival and departure. Then there was the journalism, the interviews, the articles, the photographs in which, as much as possible, AE was always to stand on the right, so her name appeared first on the caption under the picture. After that, there were personal appearances, flying in, stepping up to the microphone for a word or two and, on the subject of personal appearance, she must stop wearing those dreadful cloche hats, they did nothing for her. So she cut her bob shorter, into the slightly dishevelled boyish look that (Putnam was absolutely right about it) suited her wonderfully.

Of course, she wasn't the only young woman with wings flying around the country. There was Elinor Smith, for a start, the Flying Flapper, who knew a thing or two about publicity herself, after flying under the bridges of Manhattan's East River; and she was hardly more than a teenager when AE was in her late twenties. George gave her a call and talked over her plans; maybe he could be of some help with her ambition to start flying the bigger two or three motor planes; perhaps she could help AE, fly her from venue to venue, they might all get together and form a company to handle the business affairs of women pilots. Elinor liked the idea a lot; she wasn't rich, she couldn't finance herself, so a publicity organisation behind her would be a real advantage. She waited for news – none came; she waited for job offers – none came; she set up new flights for herself and called the press – no one arrived; she tried to interest sponsors in a big plane record – it seemed she'd been pipped at the post by the world's greatest woman flyer, who was planning her own flight.

Then there was Lady Heath, charismatic and aristocratic (at least in American eyes), a pilot with real achievements behind her: what were her plans while she was in the States, George asked, over bootleg cocktails at a dinner party held in her honour. She told him and, in the following weeks and months, her lectures and demonstration flights melted away like ice left at the bottom of the glass. And Bill Lancaster, a likeable chap with experience of big planes and a few hopes of his own for lecture tours – why didn't he just help out George and AE on those occasions when she needed to fly into a town with a posse of pressmen and secretaries and cameramen; he could be the mechanic and if he just happened to fly the trimotor plane, well, there was no need to spread it around, after all, they hadn't come to see old Bill. If AE stepped out of it after executing a perfect landing, why deprive the crowds of America's preferred aviatrix?

And finally, to get the show on the road for good, there was Sam Chapman, AE's fiancé before the Atlantic flight and afterwards . . . well, it had never been a passionate affair and he and Amelia stayed friends but, as Sam could surely appreciate, it was all really quite impossible.

What did AE think of George's schemes? It could be that she was too busy lecturing and giving demonstration flights or in writing improving letters to her mother and somewhat wayward sister and brother-in-law, another drinker, to ensure that whenever they appeared or spoke in public they didn't damage her image. She was also beginning to see herself as a representative of women in the air, and thus of women's right to achieve in any field without comparison with anyone or anything apart from their own ambitions; sex, she wrote, should be neither a hindrance nor an excuse. So, on the whole, she was probably content to leave George to do his job while she did hers; although there were occasions when she put her foot down about his relentless marketing. He had set up a line in Amelia Earhart hats for children, which were quite expensive and poorly made. When she saw them, AE refused to have anything to do with what she saw as a scheme to market shoddy goods to make money out of children. Putnam protested, but she would not budge: the whole run had to be junked.

The limit of her involvement in his myth-making was almost certainly the understandable vanity of not disagreeing when he put her accidents in landing and taking off down to mechanical causes or ground faults. She certainly never claimed to be the best woman flyer in the world and, if she didn't disagree too volubly with George's publicity machine, who's to blame her? She was absolutely certain she was carrying the ambitions of the women of America along with her on every flight.

The Powder Puff Derby of 1929 was a good place to stretch her wings. She had a new Lockheed Vega cabin monoplane and, with any luck, should stand a good chance of placing in the heavier engine class. The race also gave her the chance of getting to know most of the prominent women pilots in the country without George getting in the way. She got on well with most of them (Pancho Barnes found her a little held back – 'she was a goddam puppet!') and was able to put her talents to good use when the race authorities tried to enforce a soft sand desert landing on the second leg. She had always been good at organising things and speaking her mind, and if she sometimes went on and on, exploring every aspect of a subject twice over, this did tend to mean that her opponents, even when they were right, were often forced into compliance by the sheer persistence of her argument.

There were many who thought the Vega, with its big Wright Whirlwind engine, was too powerful a plane for AE, particularly since she'd suffered a bad ground loop a few months before, but no one ever accused her of lacking courage and she handled the ship like she'd been flying it for years. Elinor Smith, who was not in the race and had no reason to love AE, watched her landing at Cleveland to claim third place. The Vega was not an easy plane to land, it came in fast and throttling back could easily lead to a stall; AE landed badly and struggled to keep the plane from flipping; she managed it but when she climbed out of the cabin Smith could see, from her exhausted face, what a courageous effort it had been.

Third place was less than AE had hoped for, but it kept her name up there on the winners board and in the public eye: more so, in fact than the winner, Louise Thaden, who was by nature a modest woman. Perhaps the best thing to come out of the Derby, as far as AE was concerned, was the foundation of the Ninety-niners Association of Women Pilots. There had been a lot of talk about getting together and setting up some kind of organisation but it was really only AE's offer of 'doing it back east' that got it off the ground. However, her insistence on seeking full equality with male pilots who were flying overpowered pylon racers that were little more than vast engines with tiny wings – known as widow-makers – led to disagreements with Lady Heath and Elinor Smith, who did not join the organisation. They were less than impressed by AE's modest demeanour, as was Dorothy Putnam, who found herself on the end of a divorce petition soon after her husband met and fell in love with Lady Lindy.

George had begun making proposals of marriage in 1929 and was to propose six times before AE at last accepted and produced the terms and conditions document, which he seemed to find quite unexceptional. He had, after all, got what he wanted: control of the end product. What AE wanted, for her own self-esteem more than anything else, was to fly the Atlantic on her own. When she mentioned the project to her husband in the new year of 1932, he wasn't that surprised. Both Putnams had been disappointed that young Elinor Smith had recently been voted Best Woman Pilot. Maybe 1932, with an Atlantic crossing, would see AE gain the award.

Preparations were begun in secret – anything that AE was involved in tended to become a public event and George didn't want to jump the gun and let someone else sneak in first to steal the glory. A Lockheed Vega was procured and sent back to the factory to have a new engine and auxiliary tanks fitted, to give it a range of over 3000 miles. It was not an easy plane to fly, as AE knew from her Powder Puff experience. Loaded to the gunwales with extra fuel, it would be even more unsteady; on the other hand, it was dependable and had more than enough range and power for the job. Throughout the spring of 1932 she practised blind flying by instrument, which she would be certain to need; she also practised using a drift indicator, since wind would be a major factor over the open sea, and studied how the overall balance of the plane would be altered as the fuel was used up.

By 20 May, exactly five years since Lindbergh (the only solo crossing so far) had set off, everything was ready and the weather looked promising. On that same day, an article by Lady Heath appeared in *New York* magazine stating that the Atlantic would surely prove fatal to any woman pilot foolish enough to attempt it. Obviously, despite the secrecy, rumours had spread, but it was too late to turn back, as the Vega lifted off from Newfoundland and headed into a clear evening sky, with the moon shining brightly over scattered banks of cloud.

Everything went smoothly until, about three hours out, the altimeter failed. At night, with possible storms, this was bad news indeed and AE must have been tempted to turn back; she didn't, she could well have felt that her reputation, based on such flimsy achievements, would not survive. She flew on, into a severe electrical storm, which threw the Vega around the sky for about an hour. Leaving it behind, flying in clear weather once more, she saw flames spurting from the engine. Shock gave way to calm thought: in the dark the always-present exhaust flames were far more visible than in daylight; it was quite natural but, as she said later, somewhat disturbing to see her engine surrounded by a corona of fire.

Cloud began to form, getting thicker, and she decided to climb – without her altimeter – to get above it. After a few thousand feet, the controls began to get sticky and she realised that ice must be forming on the control surfaces. Before she could even think of beginning a descent, the Vega went out of control and began to fall, spinning faster and faster as it descended. She struggled to pull out, not knowing how close to the ocean she was; when she at last regained control and, by sheer muscle power, pulled the plane out of the dive and into an upward curve, she could see the white caps of the big Atlantic rollers all around her out of the cabin window. She climbed and steadied off under the cloud and flew low until she hit a fog bank, when she climbed again to what she estimated, and hoped, was a reasonable height and flew on into the darkness.

Her head began to swim. It wasn't due to tiredness or panic; it was because of the fumes of petrol leaking into the cabin from a ruptured fuel line. She began to wonder if she would actually make it and, as dawn light appeared, thin and grey, she scanned the horizon for any sight of land. At last, after 15 and a half hours, she spotted a coastline. There were no obvious landmarks but, unless she'd been blown seriously off course, it must be Ireland. She put down in the first green space that looked big enough; as soon as the Vega came to a halt she opened the cabin door to let out the stink of petrol. The morning was still and quiet – at last a figure came into sight and trudged across to her. She asked him, 'Where am I?'

'In Gallagher's Field,' he said.

She had done it, and on her own this time. Now she could call herself a pilot.

Amy Johnson's schedule in Australia was to turn out more exhausting than the flight – and more harrowing too, since the companies sponsoring her tour appeared to think they'd bought her body and soul and that she should be at their disposal day and night. When she got back home to London, things were no calmer. The *Daily Mail* had set up a whole series of appearances, at all of which they wanted her to fly in and fly out. It must have seemed like an endless haul in front of her, leaving no time for herself; she hated the idea and, as had happened before, when she had found situations becoming intolerable, she simply shut off, closed down, said that she'd collapsed with nervous tension and needed rest and quiet. And she probably wasn't that wide of the mark. The main characteristic of her appearances in public was her 'everyday' quality; she wasn't artless but she wasn't a professional personality either; she hadn't developed the skills of performing within herself; she always gave her public everything she had to give, and while that was fine for a few days, even a week or so, after a few months she was finding it impossible to rebuild her stores of energy and enthusiasm.

Her family were supportive; her father was handling the business side until she had time to give it her attention, but he had no experience of publicity (very few people in Britain at that time did); it was an American art and there was no George Putnam for Amy.

In time she began to develop a carapace, and was sometimes criticised for the icy personality she presented and the hauteur with which she approached admirers, but she was finding her own way and trying to survive as best she could in a world where she didn't know the rules. Yes, she had wanted fame of some kind, otherwise she'd never have set out on the flight to Australia, but she had no idea what the mass adulation would be like once it descended on her, simply because no one had; she was the first of her kind, not an actor or politician, an aristocrat or a soldier, she was just Amy, an extraordinary ordinary woman.

She wanted to get away, and flying seemed like the answer. She tried an England–Peking flight in 1931, but got no further than Poland; later that year she set up a London–Tokyo flight with her old friend from Stag Lane, engineer Jack Humphreys. She'd always kept a soft spot in her heart for the boys in the machine shop, just as they had for her, and one of the few speaking engagements she had kept with pleasure during the most hectic weeks of the public adulation was an invitation to attend a banquet given in her honour by the Society of Engineers. She spoke for over an hour and received a standing ovation at the end, and there's no doubt that she felt far more comfortable around engineers than she did in the Byzantine world of the society and media folk who were demanding ever-larger pieces of her.

Flying to Tokyo with Jack Humphreys, a record-breaking 4000 miles, was one of Amy's happiest experiences in the air; there was no great pressure, Jack kept the plane flying perfectly, they didn't get lost or run into any storms and if her landings were no better than usual (even Jack had to admit they left a lot to be desired) the flying itself was a pleasure and their welcome in Japan exuberant but well mannered. Though the flight did set a record, it wasn't featured that much in the

British press, mainly due to the arrival of pilot Jim Mollison in London after a record-breaking solo from Australia, beating the previous best time by two days.

Mollison, a Scot, had been flying airline routes in Australia, and had met Amy on at least a couple of occasions during her visit, when he'd flown her on scheduled passenger flights to various engagements. He was known as something of a Casanova by his airline colleagues and would certainly try his luck with any girl or woman he fancied, passenger or not. He was undoubtedly good looking, charming, a great storyteller who knew how to have a good time and what cocktails to order and just how to summon the maître d' at any hotel anywhere in the world to get the best table in the room. He had a terrific sense of purpose and self-worth and, though it is a cliché to say that small men are often aggressive and ambitious, in his case it was true. He was also a good pilot and a good sport; in his Johnson biography David Luff tells the story of Mollison's arrival at Croydon after the record-breaking flight from Australia.

Someone, assuming that the flyer was an Australian, had the bright idea of bringing a kangaroo to the celebrations. The animal got scared and hopped across the runway just as Mollison was landing. With consummate skill, he avoided the creature, which returned the favour by hopping up to him after he'd climbed out of the cockpit and kicking him in the stomach. He managed, somehow, to keep smiling through it all. He also received a telegram of congratulations from Amy Johnson, who was touring the north of England with Jack Humphreys.

The lecture series was only half completed when Amy was struck by stomach pains and whisked into hospital for a hysterectomy. There was no question of resuming the tour, so she decided to take a holiday and booked herself a cruise to Cape Town. She arrived a day before Jim Mollison flew in, on yet another record-breaking flight; he was becoming one of the big names in world aviation and it wasn't surprising that Amy should be at the airfield to welcome him on landing. Continuing his practice of eccentric arrivals,

Mollison missed the airport – some said because of exhaustion, others speculated about cocktails in the cockpit – and landed on a nearby beach, driving his Gypsy Moth down into the surf. Clambering out, he collared a taxi and arrived, soaking wet but charming and suave as ever, at the airport reception a few minutes later.

The next morning Amy spent a few hours with Jim opening telegrams, and joined him for lunch at a Cape Town restaurant. There was nothing unusual there; they had met, if briefly, before, they were both record-breaking flyers, what would be more natural than that they should spend a few hours going over the flight? And, besides, Amy's ship was leaving the next day and Jim was engaged to a society girl back in London and any self-respecting hack would immediately have smelled romance in the oleander-scented air.

The cruise liner took a meandering route back to England and by the time Amy reached home, Jim was back too; they met one afternoon, by chance, at Stag Lane airfield – Jack Humphreys, who was there, reckoned the whole thing was a set-up – but, accident or design, Jim invited her to lunch the next day at a London restaurant. She accepted. When they met, he proposed. And she accepted. Why?

David Luff, biographer of both Johnson and Mollison, who knows as much as anyone ever will about the couple, is quite certain that there was real love on Amy's part. As for Mollison, recalling his childhood experiences of a violent and alcoholic father, his parents' divorce, an RAF career flying dangerous missions in clapped-out planes over the North West Frontier (when he started to use alcohol as a support), Luff speculates that the Scot was as much in love with Amy as he was capable of being with anyone – which, unfortunately, wasn't that deeply. Almost certainly, he had learned as a child to use his undoubted charm and ability to manipulate others as a survival skill and, by the time he met Amy, the psychological pattern had become set, he couldn't have been other than he was; which was going to be unfortunate for his new wife. His society girlfriend was given the brush-off a few days later at the same London restaurant; if she thought about Mollison, or saw him at cocktail parties in the years ahead, she probably thought she'd had a lucky escape.

Amy didn't invite her parents to the wedding; Jim didn't invite his mother. Amy's parents and her sisters came anyway, though it didn't do them much good, since the wedding was over and they saw their daughter and new son-in-law for barely a moment before the smart set hurried off to the reception at a swanky London hotel. Since Amy undoubtedly loved her parents and respected them too, one can only assume that in some part of herself she felt she couldn't, in all honesty, defend the lightning courtship and sudden wedding. It was hurried, it was unwise, but she was on top of the world and loving every minute of the experience. Life was a holiday but, as she must have known, it's always a mistake to think you'll be happy if you move into your holiday home for good.

There was no settling down for the Mollisons – Amy took Jim's name, a mistake neither AE nor George would ever have made. On her Atlantic crossing Putnam had even vetoed naming AE's plane in case it detracted any attention from the pilot. But then Amy didn't like too much attention being paid to her and was happy to get back to what had become the new family business of making flights to far-away places. At first they flew apart: Jim set up a record crossing of the Atlantic east to west, the first time it had been done. Amy flew to Cape Town, breaking Jim's record and setting a new time for the return journey. Jim flew the Atlantic – again – this time from England to Brazil, yet another first. Then, when it began to look as if they would run out of records to make, they overcame their rivalry and flew together. There would be a whole slew of husband and wife firsts to rack up. They decided their first outing would be an attempt on a world long-distance record, to Baghdad and back, by way of New York.

They bought a new plane. The money was still flowing in faster than it was flowing out; although Jim, who never thought about cost, any more than he bothered about paying bills, was spending as fast as he drank. The new ship was a de Havilland Dragon, two-motor cabin biplane which they called *Seafarer*. It was adapted for the flight and thoroughly tested by both pilots. They needed to be sure of the ship since they could be certain of nothing else, certainly not of success; the Atlantic had claimed a number of flyers who had attempted east–west crossings against the prevailing winds, and the Mollisons were attempting one of the longest routes, from Wales to New York. Amy called it 'the greatest adventure of my life', when they set off for the second time – the first attempt had resulted in a broken wheel strut when the *Seafarer* hit a stone on the Welsh beach during take-off.

They shared flying time, although Jim took off and would also land. Conditions were good, though there was a headwind buffeting them continually, which naturally required more power and greater expenditure of fuel. After 29 hours they found themselves over Bridgeport, Connecticut, with their fuel tanks running dangerously low. Both were extremely tired – sleep was virtually impossible in the cramped, deafening conditions of the cabin – and the prolonged concentration required of pilot and co-pilot had resulted in an almost fugue-like state, almost certainly aided, in Jim's case, by alcohol. As he came in to land, with not enough fuel left for an overshoot and go-round, he missed the runway and hit a drainage dyke. The plane flipped on to its back into a marsh and began to sink. Trapped, held upside down in her seatbelt, Amy's head was slowly going under; she finally struggled free and clambered out of the cabin. Jim had been thrown through the windscreen and was dangling from a tree somewhere in the dark. Ground crew from Bridgeport Field were quickly on the scene, getting the aviators clear in case the plane caught fire. Local people who had been watching the landing, felt no qualms and swarmed over the plane, stripping it of everything removeable.

The Mollisons were taken to hospital. Amy's injuries were light, bruises only; Jim's exit through the front screen had resulted in a number of deep cuts but, after extensive stitching, he was up and about again. Despite the crash, they had made the crossing, the first flyers to do so from east to west, and they had no intention of missing the celebrations planned on their behalf. For Jim, that was what it was all about: the heroic aviator who could bewitch any woman with his looks, his wounds, his stories, his achievement and . . . as he must have recalled, with a sinking feeling, his wife along too. This was going to be a strictly monogamous visit.

New York gave them a tickertape parade; they loved Amy and they quite liked Jim. The couple made radio broadcasts and got an invite to the White House, along with Amelia Earhart, where they had their photos taken with the President and the first lady and the first lady of flight. AE invited Amy to stay – in fact, she invited both Mollisons, but Jim wanted to get back to the UK to supervise the readying of a replacement plane (the long-distance record was still on as far as they were concerned) and, one suspects, not to be with his wife. Because there were tensions; there were bound to be during an Atlantic flight (the couple had been seen arguing in the cockpit over Boston by the pilot of an escort plane) and, after the euphoria of the New York parade, they were making themselves apparent once more. Perhaps both felt that a little time apart would help matters; perhaps Amy was so excited by the prospect of visiting Amelia Earhart that she didn't think about it; perhaps Jim just needed a drink and the challenge of a pretty face across the room.

Amy was impressed with American attitudes towards women and flying but this may have been because the American woman she knew best was AE, who was doing significantly better than most of her sisters, while encouraging them to dare to achieve their potential in her journalism. She was on the board of at least one commercial company and could always find a market for anything she might choose to write. She was also assistant to the traffic manager of Transcontinental and Western Airlines (TWA), who were trying to encourage more travellers to get over their instinctive suspicion of planes and take to the air for their journeys. She asked Amy if she'd like to fly for TWA as guest pilot for a few days? (How dangerous can passenger flying be if we let a woman pilot the plane?) Amy jumped at the chance: TWA used Ford Trimotor carriers and this would give her a wonderful chance to get some practice on the bigger ships she would need for the long-distance flights she and Jim had in mind. There was also another aspect to her enthusiasm: she knew that sooner or later the age of record-breaking would end and she would be faced with the problem of earning a living in aviation, rather than entertainment. Any experience she could get now of working with commercial airliners and passenger schedules would, she was certain, prove valuable when it came to job-hunting a few years down the line.

Back in England she was able to put her newly learned skills into practice sooner than she expected. A Scottish chocolate manufacturer, Sir Macpherson Robertson, announced an England–Australia race worth £15 000 and a whole lot of glory to the winner. Entries were flooding in from some of the biggest names in air racing. Colonel Roscoe Turner, who usually flew with his pet lion cub, and Clyde Pangbourne, one of Bill Lancaster's character witnesses in his murder trial, were entering a Boeing 247 airliner; Sir Charles Kingsford-Smith and G. P. Taylor were in and had received criticism from the English press for racing an American Lockheed Altair, though Smith said he couldn't find a home-made ship fast enough; the young American businesswoman and flyer Jackie Cochran was flying a GeeBee Racer with co-pilot Wesley Smith, and C. W. A. Scott, three time Australia–England record-holder, had put his name down too. It was definitely the race to win and Amy and Jim felt that it could be the re-making of their relationship, if they could only find a plane good enough to stand a chance.

One was on the way – in fact, three were on the production line at the de Havilland factory at Hatfield only months after the announcement of the race. The company had decided that they would design and build three ships specifically for long-distance racing. The experience would stand them in good stead for the future development of commercial aircraft (in fact, the plane that grew out of the development model was a fighter-bomber, the Second World War Mosquito) and, if they could win, the publicity for the British aircraft industry would be priceless. They came up with a sleek two-engine design with a top speed of 220 mph and a range of nearly 3000 miles. They called it the Comet and offered it at a subsidised price of £5000.

From Amy's point of view, the race would not only give her and Jim a chance to work together again, it would also get him away from the nightclub scene where he was, by all reports, up to his old, charming, tricks again. The couple bought a Comet and had it painted black and gold. They'd both flown the route and felt they had as good a chance as anybody, though their chances of reconciliation didn't seem nearly as likely. Though she had paid half the £5000 cost, Jim was reluctant for Amy to test fly the Comet on her own. It was something of a brute to take off and land and frankly, he didn't think she had the skill to do the job. She did and told him so, and in the face of his loud opposition, told the ground crew to get the plane ready for a flight. She flew it perfectly, landing and taking off without a hitch.

Twenty planes started the race. Jim and Amy had drawn first place in the line-up and flew into clear sky. Every 45 seconds behind them, another competitor took off. There were four checkpoint stops before the winning post in Melbourne, and the Mollisons kept their lead for the first two, Baghdad and Karachi; then the problems started. The undercarriage wouldn't retract on take-off from Karachi and after struggling with it for several hours, the couple turned back to have the problem rectified. They took off again but were back within two hours. Jim hinted that they didn't want to fly on with a suspect undercarriage and maps that weren't accurate. Amy said nothing. The real reason, if there was one, remains unknown. Suffice it to say that the extra time cost them the race, which was won by Tom Campbell-Black in another Comet. The defeat did nothing for the relationship – the couple argued bitterly, loudly, constantly and in public at various airfields during the race. Whatever Amy had hoped for hadn't materialised; Jim's fear that his wife was far more popular with the general public than he was seemed to be emphasised with every passing day. He was quite prepared to share *his* affections among Amy and any number of other women, but he wanted to stand on the number one spot all by himself, and he couldn't because she was already there. It rankled and, as he had done while flying those risky missions over Afghanistan, he began to fall back on the booze, drinking not for conviviality but for need, to keep out the dark thoughts.

Amy was to make one more front-page flight in 1936, setting a record for the London–Cape Town round trip; Jim had nothing to do with the attempt and in 1938 the couple formalised their lack of relationship with a divorce.

After the Cape Town flight, Amy decided it was time to turn her experience into a real job in aviation; she was about to get a very big shock.

In January 1935 AE flew from Honolulu to California in 18 hours and 15 minutes. When she landed, a crowd of over 10 000 saluted her. One wonders if Bobbi Trout's plan to fly the same route a few years earlier had fallen victim to the George Putnam machine? Arranged or not, it was an impressive flight but not, perhaps as impressive as the 'AE flight', which had already kept the aviatrix in the air and in the forefront of the public view for a good few years. It helped that she had something to say, particularly to women's colleges and institutions, where she often lectured about 'an educational system that goes on dividing people according to their sex and goes on putting them in little feminine or masculine pigeonholes'. She once scandalised the august Daughters of the Revolution, who were agitating for rearmament in the face of the threat from Germany, by telling them that it was absurd to demand more weapons while denying women the right to use them in the field. This didactic streak had always been a part of her character, from the instructions she used to issue to her schoolmates and the strictures against her teachers, to the letters she wrote to her sister Muriel telling her how to dress her children before any public appearance and the letter she'd written to George before their marriage.

Needless to say, they had stayed together and appeared a devoted couple. George was certainly devoted to Amelia and she came to rely on him more and more, not only for the nuts and bolts of her career but also for providing the emotional ground crew to her ever-extending flights. There were rumours about both of them: that he wanted her to stop flying and have children; that she was having an affair with Paul Mantz, the ex-stunt pilot who provided technical back-up for the Honolulu flight and was taken on as the strength for a big, upcoming but so far secret record attempt the Putnams were planning.

George had met Mantz when he was in Hollywood developing projects for various studios. The stuntman was chief pilot for the film *Wings* and was happily spreading his usual romantic mayhem around the starlets and any other woman he could convince to share an illicit cocktail. He was a man in the Mollison mode and, as a film pilot, in an even more glamorous business, something which he took full advantage of. For AE, working with Mantz gave her a chance to throw off her seriousness – not that she even approached the far edge of wildness, it wasn't in her character to do so – but through him she was able to share in the 'wildness experience'. They planned business ventures together, set up cross-country flights and entered the 1935 Bendix Trophy Race, taking fifth place. Mantz's wife Myrtle was certain he was having an affair with AE; but then Myrtle thought that Mantz was having an affair with every woman he met, which was probably true (or as true as Mantz could make it) and she named AE, among others, in her divorce petition. As for whether there was anything to the stories, everything we know about AE's character, good and bad, leads to the conclusion that she would never have done anything underhand; she was simply incapable of subterfuge. Whatever the consequences, if there had been even an overnight relationship, she would have told George, because not to do so would have been unfair. The concept of the 'white lie' simply didn't exist in her world and, needless to say, George didn't believe a word of it either.

He was looking into the future. He could see that the age of the glamour pilot was coming to an end. The big airlines were carving up the world into profitable passenger routes and no one was going to come and cheer for someone who'd flown a single-seater where they'd just sat in comfort with a cup of coffee and the ministrations of a stewardess. Neither George nor AE were overly concerned about what was to come; they had business ventures and educational commitments – AE was in a development programme with Purdue Women's College, among many other projects. There was, however, the chance of one last great flight which would, Putnam was convinced, indelibly stamp AE as the greatest woman pilot in history. Once she'd been round the world, there would be no more major records, no more heroics on a grand scale; the curtain would come down – *fin d'opéra*.

What were AE's feelings? She knew she'd need a bigger plane, a two-engine at least; she'd also need a navigator/ mechanic aboard, since she'd be pushing herself harder than she'd ever done before – than anyone had ever done before. There had been round-the-world flights: Wiley Post was the first in his plane *Winnie Mae*, but no one had yet flown the equator, a round trip that would cover almost 30 000 miles and put a huge strain on crew and machine. It could be done, AE was convinced of that, and she'd racked up a lot of experience since that first passenger flight across the Atlantic, but was she the one? George thought so, and after they'd talked it through, so did she.

She needed the right plane. And they found that in the Lockheed Electra, a sleek two-motor passenger carrier. George pulled in a number of business backers and AE had little trouble in persuading the authorities at Purdue College to adopt and finance the plane as a flying laboratory. Various trade magazines criticised both her and the university, saying that the laboratory plane was no more than a gimmick, (and perhaps it was) but, from an educational point of view, it was also a statement about female aspirations.

AE entered the 1936 Bendix, this time flying with co-pilot Helen Richey, who had experience of multi-engine planes; Paul Mantz was again part of the team, providing back-up on the ground, as he would be during the world flight. AE didn't place – Powder Puff winner Louise Thaden took the trophy – but she did begin to get to know the characteristics of the Electra.

The first choice of navigator was the immensely experienced Harvard geographer Henry Washburn Jr. The flight was to head west and, looking over the maps, Washburn immediately homed in on Howland Island to be the second stop after Honolulu. It was a tiny speck of land virtually in the middle of the Pacific. He asked how AE intended to find it. She said, by dead reckoning. Even with his experience, Washburn said that he wouldn't trust himself to hit so small a target after 2000 miles across unmarked ocean. He suggested installing a homing beacon on the island, but neither AE nor George liked the idea and Washburn found himself off the shortlist. The final choice was Fred Noonan, a pilot and navigator for PanAm's Pacific service, who had more hands-on experience than Washburn; he was also a man who liked a drink (as many flyers did) but AE accepted his assurances that he wouldn't drink too much during the trip. One wonders why, given her experience with her father and *Friendship* pilot Wilmer Stulz, she was so confident.

The Electra, with AE, Noonan and an extra navigator, sea captain Harry Manning aboard, took off from Oakland California on 17 March 1937. They experienced no problems on the Honolulu flight, arriving after 16 hours. The next leg was the critical 2000 miles to Howland Island, but the Electra experienced a technical failure on take-off and the undercarriage and one engine were smashed up. Paul Mantz, who observed the mishap, was convinced that the crash occurred not through technical problems but because of AE's bad take-off habits, using the engines to turn the ship rather than the rudders. Either way, the Electra had to be dismantled and flown back to the Lockheed works for rebuilding, since there was no thought of cancelling the flight; too many commercial deals were in place for that and, besides, AE did not like the idea of anyone thinking she was looking for a way to chicken out. Sailor Harry Manning declined to sign on for another voyage (AE was too bull-headed for him) but Fred Noonan was happy to stay with the team.

The second attempt would reverse the flight direction. Due to weather changes they decided to head east, leaving the problem of finding Howland until the penultimate leg. By then, AE and Noonan would have clicked as a team and their navigation-instrument flying skills would have been honed by hundreds of hours in the air. They would also be nearing their physical limits, but no one seems to have thought of this: AE had always been able to find that little bit extra, pushing herself until, as in France in 1918, she could push no further. But now, of course, there was always George to give her that last little shove.

The rebuilt Electra was delivered to Miami. AE told the press that she felt she had one more good flight left in her. Noonan wrote to his wife that there was no woman pilot with whom he would rather fly. The plane was checked, but not by Paul Mantz (rumours were that he had been intentionally excluded, presumably due to his criticism of AE's flying skills) and they set off on 1 June 1937. One of the things they left behind, that Mantz would have insisted was in place, was the 250-foot trailing radio aerial which was a constant problem to use but which increased the radio range of the aircraft by a vast distance.

They went down to Brazil, east across the Atlantic to Dakar, to Karachi, Calcutta and Rangoon, down through Indonesia to Darwin and across to New Guinea, making 28 stops in all and covering 22 000 miles. Before them lay the Pacific and, 2556 miles away, Howland Island. Through government contacts (Secretary of Air Gene Vidal was a good friend and business partner, FDR and Eleanor were following the flight with interest) it had been arranged that a US Navy ship would be in the vicinity of the tiny island, broadcasting on agreed wavelengths, to help the Electra locate its target. Of course, without the trailing aerial, that was going to be a good deal more difficult than planned but the Electra's crew had no doubt they could do it.

AE delayed their take-off from New Guinea by a day, which was just as well since Fred Noonan had got drunk as a skunk on the evening of their arrival. Apparently he was annoyed he had not been included in an invitation to a celebration dinner held by local bigwigs. It's more likely that, exhausted after a month of hard flying, with the hardest stage yet to come, he simply felt like trying one on. There's no doubt that AE was equally happy to wait the extra day; she must have been very near the limit of her own endurance and the chance to rest even for a few hours, before attempting the Howland flight, would have made real practical sense.

On the morning of 10 July they refuelled the Electra, putting in enough fuel to give them a few hours' leeway, should they need it, though Noonan was confident he would locate the island with no trouble. All extra weight was removed from the ship, the crew climbed aboard, AE admitting that she'd be glad when the next leg was behind them, and at 10.00 a.m. they took off.

As they flew west into the night and the following day, AE sent a number of radio messages stating their position and direction; all seemed to be going well. At dawn on 11 July she sent: 'Am about 100 miles out (from Howland)'. The listeners and watchers on board US Coastguard cutter *Itasca* expected to see the Electra any moment. Nothing appeared in the sky. An hour and a half later there was another call, loud and clear, saying that AE couldn't see the *Itasca*, though she should be in view. She added that fuel was getting low. *Itasca* started making smoke, which should have been clearly visible for miles in the cloudless Pacific sky. Still no response. Twenty minutes later, AE came through with: 'Circling but have lost your signal'. Then, moments later, after *Itasca* sent with everything she had, AE came back to say she'd relocated the signal.

There was nothing more for over an hour, then the plane's signal came through loud and clear. AE gave the call sign: 'KHAQQ to *Itasca*. We are on the line of position one five seven dash three three seven. Will repeat this message on 6210 kilocycles. We are running north and south.' According to the radioman on duty, her voice sounded frantic.

There were no more messages.

A search was organised immediately; the President himself commanded that no effort be spared. Nothing was found: no wreckage, no oil slick, absolutely nothing. It was as if AE and Fred Noonan had vanished into the thin air of legend. Theories sprang up by the dozen over the next few decades. She was spying on Japanese war plans for the government and shot down; she crash-landed on a secret Japanese arms dump and was executed; they landed on a deserted island and died of starvation. The most likely explanation came from historian Roy Nesbit, who thinks that a deeply tired Noonan made a mistake in his measurements on the morning of 11 July, calculating their position at sunrise from sea level rather than their actual height above it. This would have put their final position, as they flew a corridor looking for Howland, 35 miles west of where they should have been.

The best explanation was given by pilot Jackie Cochran at a memorial for AE when she said that there had been no winning or losing, for AE's last flight was endless . . .

Sitting in a rented cottage in the village of Haddenham, Buckinghamshire, England, Amy must have experienced many emotions on hearing of AE's death. She had finally decided to move out of London. While Jim Mollison was still flying, setting whatever records he could still pick up, Amy's position tended to be seen as the deserting ex-wife. The press is not particularly forgiving of those who stop doing its bidding and no longer care to live a front-page life. Even so, the early news of AE's circumnavigation had awoken thoughts of one last great flight of her own; unfortunately she no longer had anything like the resources she would need, and no one to help her find them. The money was running out; that was another reason for country living. But she was happier away from the nightclubs and press conferences, the dinners and dances. She was able to ride, something she came to love, and to find a new kind of pleasure in the air through gliding.

The greatest contrast between her own quiet life and AE's bandwagon was still the material question. Never mind setting world-spanning records, Amy had found it impossible to get any kind of decent job in the British aviation industry. She had found a temporary position with Hillman airways (later BA) flying a de Havilland Dragon, although there were always some passengers who refused to go up with a woman pilot. The job only lasted a few months, since she was preparing for the MacRobertson England–Australia race, and when she tried to get another commercial position after the event, she found there was nothing available. She wrote articles for magazines, appeared at the odd trade show and was aviation editor of the *Daily Mail*, commuting from Haddenham when necessary. She also indulged in a new sport, rallying with a friend, competing in the 1939 Monte Carlo rally.

Her main interest and hope for a future in which she could do something useful for flying and the country, lay in the international situation. Anyone with a brain, except the majority of the Conservative government, could see that war was inevitable; and anyone who knew anything about flying realised that airpower was going to play a vital part in the coming struggle. The government had started a scheme to train part-time pilots to act in a civil defence capacity and were looking for instructors. Amy thought she would be ideal and, calling in a few favours, was able to get an interview with the Minister for

Air, Sir Kingsley Wood. He listened to her plans and offered her a job as a clerk. It was nothing short of a calculated insult and, given her experience and her ability to get on with people, not far short of stupid either.

The war did mean there were other opportunities. Small commercial companies were ferrying materials and men around the country and they needed pilots. So Amy got back in the air as a glorified taxi driver. She didn't mind; at least she was doing something, no matter how small. Then, in the first, oddly quiet months of the war, another chance came along. The Air Transport Auxiliary (ATA) had been set up to train older, commercial, male pilots to deliver military planes to RAF bases, should hostilities arise. British women pilots were not long in drawing attention to the unfairness, not to say the inefficiency, of an all-male transport service. Why shouldn't women pilots fly aircraft from the factory to the airfield just as well as men? There was no reason at all, the authorities decided, and started to recruit women pilots for the service. The skills were there, the flyers were ready but above all they needed a leader, a figure who could inspire the new recruits and fight for the service in the trickiest war of all, that of government bureaucracy. Amy Johnson seemed a natural for the job; and naturally she didn't get it. Fortunately for the service, it went to the talented 29-year-old Pauline Gower, not only because of her wide experience, natural ability and her MP father, but also because she wasn't Amy Johnson. She had never been, as one government minute put it, a *stunt pilot*.

Gower immediately asked Amy to join the first cadre of ATA women pilots but a little time needed to pass so that hurt vanity might be soothed before Second Officer Johnson took up her piloting duties. Once she did so, she proved, as always, hugely popular at every airfield she visited and she still managed to fit in with the far less experienced women working alongside her. She also found herself working with her ex-husband, who was too old for the RAF and had joined the ATA. In many ways it was the remaking of him. He gave up drinking, worked responsibly and was generally well liked by everyone; he was still a fine pilot and cool enough to escape, in an unarmed air taxi, when he was attacked by a German fighter. After landing and inspecting the bullet holes in his fuselage, his greatest desire was to have a cup of tea! Whenever she and Jim met, he would ask Amy out on a date, but he was still married to somebody somewhere, no longer to her, and she always said no. He sent her a pair of slippers for Christmas.

On 5 January 1941 Amy Johnson climbed aboard an Airspeed Oxford she was due to deliver. The weather at the Blackpool airfield was bad. A thick mist had drifted in off the North Sea and a drizzling rain cut visibility even further. ATA pilots on the station were advised to wait for clear weather or even postpone their flights until the Monday. Amy decided not to. The last anyone ever saw of her was her take-off that morning. She had told one of the riggers that she intended to get above the cloud and into the sunlight . . .

Sources for each chapter

Clover Field. The Bobbi Trout interview brought the time and place to life; Gene Norah Gessen's *The Powder Puff Derby of 1929* tells the story of the race from start to finish and follows the subsequent careers of the participants; *Race with the Wind* by Birch Matthews; the writings and newspaper columns of Will Rogers; *The Air Racers* by Terry Gwynn-Jones; *Women Aloft* by Valerie Moolman; *Through the Overcast* by Assen Jordanoff; *High Wide and Frightened* by Louise Thaden and the Ninety-Nines website also provided material.

Print the legend. The Bobbi Trout interview provided much of the information about Pancho's flying in the late 1920s and 1930s. For the early years and her career after the war I relied upon *The Aeronauts* by Donald Dale Jackson and *Women Aloft* by Valerie Moolman. *Pancho, The Biography of Florence Lowe Barnes* by Barbara Hunter Schultz; *The Happy Bottom Riding Club* by Lauren Kessler and Grover Ted Tate's *The Lady Who Tamed Pegasus* are all invaluable accounts of a fascinating woman.

Bicycle dreams came from many sources: *Flight* and *Aeroplane* magazines for the relevant years, newspaper reports from the *Daily Mail* and *The Times*; *First in Flight, the Wright Brothers in North Carolina* by Stephan Kirk; *The Wright Brothers: A Biography*, Fred C. Kelly; *The Papers of Orville and Wilbur Wright* edited by Marvin W. McFarland; *This Flying Game* by Arnold and Eaker; *The World Encyclopedia of Civil Aircraft* by Enzo Angelucci; the Bobbi Trout interview; *The Spirit*, Brooklands Museum magazine, gave information on Hilda Hewlett; *Before Amelia* by Eileen F. Lebow covers the period and the personalities in an exemplary fashion; *Katherine Stinson, The Flying Schoolgirl* by Debra L. Winegarten; *Flying Start*, M. H. Goodall's history of early flying at Brooklands; *The Sky's the Limit* by Wendy Boase; *The Bedfordshire Magazine* on the Hewlett factory; the history of aviation website of Monash University; the National Air and Space Museum website and Mike and Dorina Graham's Melli Beese website.

Death of an unknown woman. There is comparatively little material available on Lady Mary Heath. Her own book, *Women and Flying*, written with Stella Wolfe-Murray, was invaluable, as were conversations with Ann Tilbury-Harrington and unpublished letters in her possession from Elinor Smith; *Throttle*

Full Open: A Life of Lady Bailey, Irish Aviatrix by Jane Falloon gives a complete and entertaining account of the life of Lady Bailey; Mary Cadogan's *Women With Wings* quotes the *Schooldays* magazine article, Wendy Boase's *The Sky's the Limit* fills in some of the good years and an interview Lady Heath gave to *The People* newspaper, and court reports from *The Times* fill in some of the bad. *Flight* magazine provides a record of Lady Heath's flying career and also printed many of her articles and letters; *Amelia Earhart's Daughters* by Haynsworth and Toomey gives the US ATA story.

Evelyn bobs her hair draws mainly on interviews with Bobbi Trout and the Trout archives maintained by Cheryl Baker; *Just Plane Crazy* by Donna Veca and Skip Mazzio underlines memory with research and documentation; *Women Aloft* by Valerie Moolman provides an overview of the times, as does Gene Norah Gessen's *Powder Puff Derby of 1929*.

'They say I shot a man and killed him' owes much to Ralph Barker, who wrote the definitive Bill Lancaster biography, *Verdict on a Lost Flyer*, and also made available extensive interviews, correspondence and personal reminiscence of Chubbie Miller. *The Air Racers* by Terry Gwynn-Jones also provided information.

Degrees of difficulty. There is comparatively little published on Bessie Coleman. Doris L. Rich's *Queen Bess, Daredevil Aviator* is a major source and a fascinating read; The Ninety-Nines, The Aviation Pioneers, The Black Wings, The Texas History and Luke Irvin's Bessie Coleman websites all give much useful information about the aviatrix and the society in which she fought to fly. Hanna Reitsch's *Flying is my Life*; James P. O'Donnell's *The Berlin Bunker*, Judy Lomax's *Hanna Reitsch*; Valerie Moolman's *Women Aloft* and Dennis L. Piszkiewicz's well-researched biography *From Nazi Test Pilot To Hitler's Bunker* and the Aviation Pioneers website present the many sides of Hanna Reitsch's eventful and sometimes shocking life.

Always stand on the right. Assen Jordanoff's *Above the Overcast* provided technical background on contemporary flying and weather science. There are almost too many biographies of Amelia Earhart to list but among those that were particularly useful were *Amelia, A Life of the Aviation Legend* by Goldstein and Dillon; *Amelia Earhart: a Biography*, Doris L. Rich; *Straight on Till*

Morning and *The Sound of Wings*, both by Mary S. Lovell, the latter telling the tale of G. P. Putnam's western adventures; *The Fun of It* and *20 Hrs, 40Min: Our flight in the* Friendship, both by Amelia Earhart; the Bobbi Trout interview; *Hollywood Pilot, The Life of Paul Mantz* by Don Dwiggins. There are far too few biographies of Amy Johnson: superseding all others is David Luff's magisterial *Amy Johnson, Enigma in the Sky*; Constance Babbington-Smith's *Amy Johnson* is the best of the earlier books; *Myself When Young* and *Sky Roads of the World*, both by Amy Johnson herself, provide background; general information on the period came from *Amelia Earhart's Daughters*, Haynsworth and Toomey; *The Sky's the Limit* by Wendy Boase; *The Air Racers* by Terry Gwynn-Jones; *Women With Wings* by Mary Cadogan; and the collection and research website of the Science Museum, London.

Select bibliography

Arnold, Hap and Eaker, Ira. *This Flying Game*. Funk and Wagnalls, 1936.

Barker, Ralph. *Verdict on a Lost Airman*. Harrap, 1969.

Boase, Wendy. *The Sky's The Limit*. Osprey, 1979.

Cadogan, Mary. *Women With Wings*. Macmillan, 1992.

Falloon, Jane. *Throttle Full Open*. Lilliput Press, 1999.

Earhart, Amelia. *The Fun of It*. Academy Chicago Publishers, 1977.

Earhart, Amelia. *20 Hrs, 40Min: Our Flight in the Friendship*. G. P. Putnam's Sons, 1928.

Goldstein, Donald and Dillon, Katherine. *Amelia, A Life of the Aviation Legend*. Brassey's, 1997.

Gwynn-Jones, Terry. *The Air Racers*. Guild Publishing, 1989.

Haynsworth, Leslie and Toomey, David. *Amelia Earhart's Daughters*. Perrenial, 1998.

Heath, Lady Mary and Murray, Stella Wolfe. *Women and Flying*. John Long, 1929.

Howard, Fred. *Wilbur and Orville: A Biography of the Wright Brothers*. Ballantyne, 1988.

Jackson, Donald Dale. *The Aeronauts*.Time–Life Books, 1981.

Jessen, Gene Norah. *The Powder Puff Derby of 1929*. Sourcebooks, 2002.

Johnson, Amy. *Myself When Young*. Frederick Muller, 1938.

Johnson, Amy. *Skyroads of the World*. W & R Chambers, 1939.

Jordanoff, Assen. *Through the Overcast*. Funk & Wagnalls, 1938.

Kelly, Fred C. *The Wright Brothers: A Biography*. Dover, 1991.

Kessler, Lauren. *The Happy Bottom Riding Club*. Random House, 2000.

Kirk, Stephen. *First In Flight: The Wright Brothers in North Carolina*. John F. Blair Publications, 1995.

Lebow, Eileen F. *Before Amelia*. Brassey's, 2002.

Lomax, Judy. *Hanna Reitsch*. John Murray, 1988.

Lovell, Mary S. *The Sound of Wings*. Hutchinson, 1989.

Lovell, Mary S. *Straight on Till Morning*. Arena, 1988.

Luff, David. *Amy Johnson, Enigma in the Sky*. Airlife, 2002.

Markham, Beryl. *West with the Night*. North Point Press, 1942.

Matthews, Birch. *Race With The Wind*. MBI Publishing Company, 2001.

Moolman, Valerie. *Women Aloft*. Time–Life Books, 1981.

Rich, Doris L. *Queen Bess, Daredevil Aviator*. Smithsonian Institute Press, 1993.

Piszkiewicz, Dennis. *From Nazi Test Pilot to Hitler's Bunker*. Praeger, 1997.

Reitsch, Hanna. *The Sky My Kingdom*. Bodley Head, 1955.

Schultz, Barbara Hunter. *Pancho, the Biography of Florence Lowe Barnes*. Little Buttes Publishing Co, 1996.

286

Sterling, Bryan. *The Best of Will Rogers*. Evans & Company Inc, 1979.

Shepherd, Dolly. *When the 'Chute Went Up*. Robert Hale, 1974.

Tate, Grover Ted. *The Lady Who Tamed Pegasus*. Maverick Publications, 1984.

Thaden, Louise. *High, Wide and Frightened*. Air Facts Inc, 1973.

Veca, Donna and Mazzio, Skip. *Just Plane Crazy*. Osborne Publications, 1987.

Weingarten, Debra. L. *Katherine Stinson, The Flying Schoolgirl*. Eakin Press, 2000.

Websites

Aviation Pioneers: An Anthology. An invaluable site.
www.ctie.monash.edu.au/hargrave/pioneers.html

The Ninety-Nines Organisation of Women Pilots
www.ninety-nines.org/

Smithsonian National Air and Space Museum
http://www.nasm.si.edu/

The Science Museum, Kensington
www.science-museum.org.uk/